JOURNAL OF
MACHINE TO MACHINE
COMMUNICATIONS

Volume 1, No. 2 (May 2014)

JOURNAL OF MACHINE TO MACHINE COMMUNICATIONS

Editor-in-Chief
Johnson I Agbinya, La Trobe University, Australia

Editorial Board
Karim Djouani, Creitel, France
M. J. E. Salami, International Islamic University Malaysia
Mari Carmen Aguayo Torres, Spain
Théophile K. Dagba, Universite d' Abomey-Calavi, Benin Republic
Semiyou A. Adedjouma, Universite d'Abomey-Calavi, Benin Republic
Frank Jiang, ADFA, Australia
Atsushi Ito, KDDI, Japan
Mjumo Mzyece, Tshwane University of Technology, South Africa
Jose Roberto de Almeida Amazonas, University of Sao Paulo, Brazil

Aims and Scope

Currently significant research publications in machine to machine communications, Internet of things, distributed and ubiquitous computing are spread across many disciplines and journals with each one focusing on a narrow aspect of the field of Machine to Machine Intelligence. Consequently finding a single point of focus which makes it easy for researchers and in particular early career researchers to latch onto new publications in this field in one place has been difficult. The objective of Journal of Machine to Machine Communications is to provide such a focus for state of the art research findings and development is machine to machine systems. The Journal will receive contributions which deal with all aspects of machine to machine (M2M) communications, Internet of things (IoT), distributed and ubiquitous computing, communications and sensing. Towards that end the team of international editors and advisory board is formed to include individuals with extensive expertise in at least one aspect of the Journal spread. The Journal is published by the Association of International Scientists (AIS).

Published, sold and distributed by:
River Publishers
Niels Jernes Vej 10
9220 Aalborg Ø
Denmark

Tel.: +45369953197
www.riverpublishers.com

Journal of Machine to Machine Communications is published three times a year.
Publication programme, 2014: Volume 1 (3 issues)

ISSN:2246–137X (Print Version)
ISBN:978-87-93102-98-9 (this issue)

JOURNAL OF MACHINE TO MACHINE COMMUNICATIONS

Volume 1, No. 2 (May 2014)

Three Dimensional EEG Model and Analysis of Correlation between Sub Band for Right and Left Frontal Brainwave for Brain Balancing Application

N. Fuad[1] and M. N. Taib[2]

[1]Department of Computer Engineering, Faculty of Electrical and Electronic Engineering, Universiti Tun Hussein Onn Malaysia, 86400 Johore, Malaysia
[2]Faculty of Electrical Engineering, Universiti Teknologi MARA, 40450 Selangor, Malaysia

Received 15 April 2013; Accepted 18 May 2014
Publication 4 August 2014

Abstract

This paper presents power spectral density (PSD) characteristics extracted from three-dimensional (3D) electroencephalogram (EEG) models in brain balancing application. There were 50 healthy subjects contributed the EEG dataset. Development of 3D models involves pre-processing of raw EEG signals and construction of spectrogram images. The resultant images which are two-dimensional (2D) were constructed via Short Time Fourier Transform (STFT). Optimization, color conversion, gradient and mesh algorithms are image processing techniques have been implemented. Then, maximum PSD values were extracted as features and further analyzed using Pearson correlation. Results indicate that the proposed maximum PSD from 3D EEG model were able to distinguish the different levels of brain balancing indexes.

Keywords: power spectral density, 3D EEG model, brain balancing.

Journal of Machine to Machine Communications, Vol. 1 , 91–106.
doi: 10.13052/jmmc2246-137X.121

1 Introduction

A normal human brain contains a hundred billions of neurons as have been figured out by the scientists. About 250,000 neurons are connected to a single neuron. The information will be processed and sent by a normal brain to whole human body. An electrical power will be generated and this signal is named wave [1–4].Brain is consisted of pair parts known as left hemisphere and right hemisphere. The language, arithmetic, analysis and speech are performed in the left side of the brain. The right side of hemisphere is dominant in the cognitive tasks such as understanding, emotion, perceiving, remembering and thinking [5–8].

The happiness and good health is affected by healthy lifestyle [9]. Referring to a psychiatrist, Dr. Paul Sorgi, the stress feeling and faces mental illness is caused by disability of mind balance control and imbalance lifestyles will be affected by physical and psychology [11]. In contrast, the happiness, satisfaction, healthy and free to communicate with each other are achieved by manage the mind balance [10–12]. Many studies proved that longer and healthier life can be obtained to ensure the human being live in balance in order to improve human potential. Recently, the interests to find the methods for balancing of the brain have been increased [13–15] by using auditory and visual methods in brainwave entrainment that results in more waves that are similar to the frequency following response [14–16]. There are other methods to perform the test namely Transcranial Magnetic or Electric Stimulation. This traditional method included massages, meditation and acupunctures [13–15]. From the previous researches and the review of literature, most of the human want to feel happy and healthy. While, a balance life is become from balance thinking or mind from the brain [1, 17]. Nowadays, there is not found a scientific prove of brainwave balancing index using EEG. But there are some techniques or devices to help human felling clam and brain balancing.

The electroencephalogram (EEG) is a device to collect brainwave signal and the frequency of theta-θ, delta-δ, alpha-α and beta-β bands are produced [19]. The EEG raw data is produced in spectral pattern. The power for each spectral powers has the frequency bands: theta-θ (4–8 Hz), delta-δ (0.5–4 Hz), alpha-α (8–13 Hz) and beta-β (13–30 Hz) [20]. These components are utilized and hypothesized to produce the variation of neuronal assemblies [1, 21]. Referring to the theory, beta band is the lowest amplitude but the highest frequency band while delta band is opposite to beta band. High beta is occurred when human is inactive, not busy or anxious thinking but the low beta is occurred in positive situations. Human activities such as closing the eyes, relax/reflecting mode and all activities with inhibition control are affected by alpha band. The theta band is occurred when human in stress mode and light sleep also it has been found in baby activities. When human is in profound sleep mode, the delta band is produced [3]. However, EEG topography is produced by several software or toolboxes such as EEGLAB in Matlab embedded module[22]. LORETA is an electromagnet tomography in low resolution and Alzheimer patients need the EEG topography [23]. MEG and EEG signal are normally displayed by using the brainstorm approach [24].

Normally, EEG signals are represented by time domain and the plot of domain is displayed in time-amplitude. In the same time, some additional information can be found from frequency domain signal. So, the method namely Fourier Transform (FT) is implemented to produce this domain. The artifact in EEG can be re-referenced in average of EEG power density spectrum analysis. The result is analyzed using an algorithm of Fourier Transform (FT) algorithm [25]. Discrete Fourier Transform (FFT) is used to estimate the smoothed periodograms by the power spectral density [26]. There are several methods to perform time-frequency analysis and Short Time Fourier Transform (STFT) is one of the method to produced two dimension (2D) EEG outcome named 2D EEG image [27]. However, some differences are recognized among 3D and 2D in term of implementation in technology field. For examples, parameters for 2D baby scanning are height and width and 3D baby scanning are height, width and depth [28]. There are another research done in 3D implementation such as crystal surfaces [29], brain-computer inter-face (BCI) [30] and assessment some parameters for 3D acoustic scattering; constant, linear and quadratic [31].

In this paper, some methods are proposed to produce 3D EEG model. The resultant of 3D model for EEG is shown and the results used to find the correlation between left and right brainwaves using features extraction of Max PSD from 3D model. The normality is tested using Shapiro-Wilk and Pearson Correlation in Statistical Package for Social Science (SPSS).

2 Methodology

The flow diagram in Figure 1 shows the methodology of the research. Some processes have been carried out; data collection, signal pre-processing, 2D and 3D development, features extraction and data analysis on maximum PSD for evaluation.

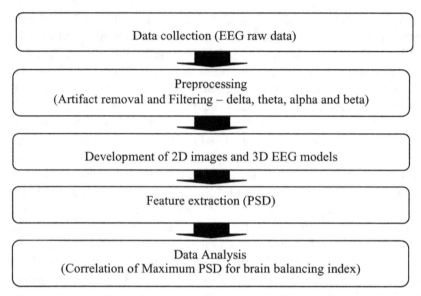

Figure 1   Flow diagram of methodology

2.1 Data Collection

This research involved 51 volunteers of samples which are 28 males and 23 females with an average age of 21.7. The data are collected from Biomedical Research and Development Laboratory for Human Potential, Faculty of Electrical Engineering, Universiti Teknologi MARA (UiTM) Malaysia. All volunteers are healthy and not on any medication before the tests. These are performed and have fulfilled the requirement provided by ethics committee from UiTM.

Figure 2 shows the experimental setup for EEG recording. The EEG data were recorded using 2-channels (gold disk bipolar electrode) and a reference to two earlobes. The electrodes connections comply to 10/20 International system with the sampling rate of 256Hz. Channel 1 positive was connected to the right hand side (RHS), Fp_2. The left hand side (LHS), Fp_1 was connected to channel 2 positive. Fp_z is the point at the center of forehead declared as reference point. MOBIlab was used in wireless EEG equipment and the EEG signal was monitored for five minutes. The Z-checker equipment was used to maintain the impedance to below than 5kΩ. The MATLAB and SIMULINK are used to process the data with the intelligent signal processing technique.

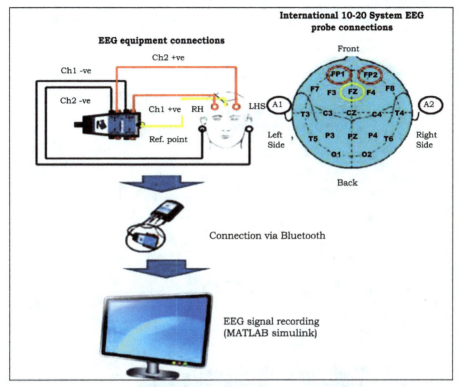

Figure 2 Experimental setup

2.2 Pre-processing

The EEG raw data was processed separately after data collection. The filter of band pass and artifact removal was included in EEG signal pre-processing. The artefacts may be produced when the eyes of volunteers blink. The artefacts can be removed by setting a threshold value in MATLAB tools. The setting of threshold values were below than -100μV and greater than 100μV. Only the meaningful and informatics signal were occurred within -100μV to 100μV. The Hamming windows was used to design the band pass filter with the rate of overlapping of 50% for the frequency 0.5Hz to 30Hz which were theta-θ (4–8 Hz), delta-δ (0.5–4 Hz), alpha-α (8–13 Hz) and beta-β (13–30 Hz). An example of raw EEG signal showed in Figure 3. This signal is from Fp2 and gain maximum value 180 μV. Figure 4 showed filtered signal using band pass filter with 256 Hz frequency sampling and the signal is in time domain plot.

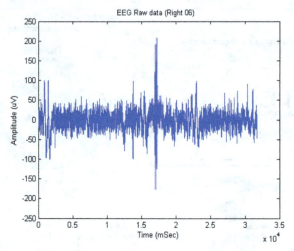

Figure 3   Raw EEG signal in time domain

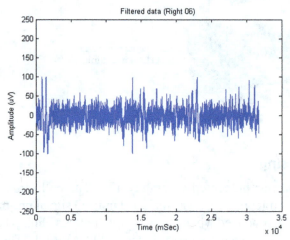

Figure 4   EEG signal after implemented band pass filter

2.3 2D Images Using STFT

The STFT was used to produce the spectrogram image in 436x342 pixels of image size for Fp1 and Fp2 channel. Each band of frequency was set in a spectrogram image. The Beta band was set from 13Hz to 30Hz, Delta band was set from 0.5Hz to 4Hz, Alpha band (8Hz to 13Hz) and Theta band (4Hz to 8Hz). This method was used for motor imagery EEG signal classification [22, 23] and detection of epileptic seizures in EEG collection [24, 25]. Therefore, the

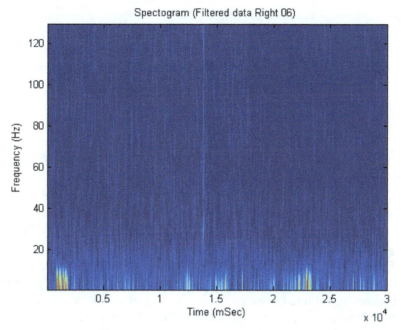

Figure 5   2D EEG image or spectrogram

analysis of time frequency (Equation 1) using STFT was performed. The EEG signal, x(t), the window function, w(t) and signiture of complex conjugate, * are stated in STFT. The signal changed in time and performed using STFT. The small window of data in one time was used to map the signal to 2D function of time and frequency. Then the Fourier Transform (FT) would be multiplied with window function to yield the STFT.

$$STFT_x^{(w)}(t, f) = \int_{-\infty}^{\infty} [x(t).(t - t').e^{-j2\pi ft} \, dt] \qquad (1)$$

2D EEG image named spectrogram is in time frequency domain. This image is generated using STFT and the algorithm has explained previously. The outcome showed in Figure 5.

2.4 3D EEG Models

3D EEG models have been developed from EEG spectrogram using image processing techniques. Color conversion, gradient, optimization and mesh algorithms were integrated to developed this model, while the spectogram images are represented in RedGreenBlue (RGB) color. Color conversion was implemented to transform spectogram of RGB to spectogram of gray scale. Gray scale images were used in a data matrix (I) which the values represent intensity within some range which are 0 (black) and 255 (white). Gray scale is the most commonly used images within the context of image processing. Equation 2 is implemented to RGB values of the pixels in the image to gray scale values of pixels.

$$P = C \times R \tag{2}$$

where C is the column value of the pixel, R is the row value and P is gray value.

Then, Optimization Options Reference (OOR) was implemented to gray scale pixels image for optimization technique. There were severals options in OOR using MATLAB software but for this research, DiffMaxChange (Maximum change in variables for finite differencing) option have been chosen. The natural shape can be found from pixels value. This shape related to the maximum of certain energy function computed from the surface position and squared norm. A finite number of points were generated for the height of the optimized surface. Then the matrices of pixels value were resized using Gradient and Mesh algorithm into vectors. Two vector arguments replaced the first two matrix arguments, length(x) = n and length(y) = m where [m, n] = size (z). A vectors x is included matrix X (rows) and a vectors y is for matrix Y (columns). Matrix X and Y can be evaluated using MATLAB's array mathematics features. The pixels value for one part of gray scale for gray scale spectrogram is shown in Figure 6. The outcome will implement optimization technique named Optimization Options Reference (OOR).

The 3D EEG model for 2D EEG image (Figure 6) is shown in Figure 7. The model has been implemented Mesh and Gradient algorithm.

2.5 EEG Analysis

A spectral of power spectral density (PSD) was produced from Three Dimension (3D) model, then the max PSD was choosed as features to analyze. Using Shapiro-Wilk technique in Statistical Package for Social Science (SPSS) software, the normality is tested. Shapiro-Wilk is selected because of the small size of samples. If the value of p is small enough which is less than 0.05 ($p < 0.05$), the data is considered as significant but not in normal distribution. Pearson Correlation showed the correlation between sub band for left and right brainwaves. Brainwave correlation is calculated using the formula as shown by (Equation 3).

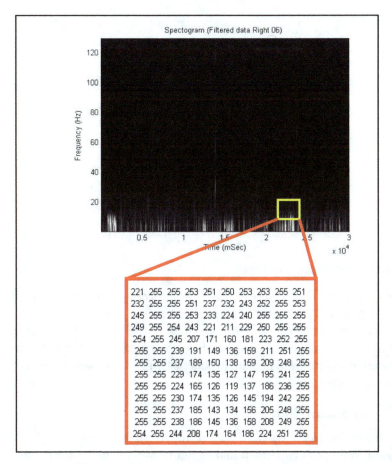

Figure 6   Pixels value of gray scale spectrogram

$$Pearson\_Correlation = \frac{\sum(x_i - x)(y_i - y)}{(N-1)s_x s_y} \tag{3}$$

where the mean of the sample is represent by and and xi and yi is the data point and N is the number of samples. Correlation is the linear relationship between two variables. Zero correlation indicates that there is no relationship between the variables. Correlation of negative 1 indicates a perfect negative correlation, meaning that as one variable goes up, the other goes down. Correlation of positive 1 indicates a perfect positive correlation, meaning that both variables move in the same direction together.

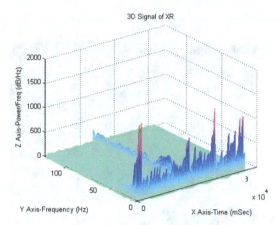

Figure 7   The 3D EEG model

3 Results and Discussion

The development of 3D EEG models have been successful using optimization; gradient and mesh algorithms as shown in Figure 8 (a)-(h) . These show each of frequency bands for Fp1 and Fp2 channels. The 3D model is spectral of PSD and a different max PSD produced by each frequency band. Eight 3D models for channels Fp1 and Fp2 are produced by EEG sample. The 3D model produced as depicted in Table 1.

3.1 Brain Balancing Index

The brain balancing index was analyzed offline from previous work [18]. The percentage difference between left and right brainwaves was calculated from PSD values of EEG signals using the asymmetry formula as shown by (2). Table 2 shows the respective index and range of balance score. There were three groups; index 3 (moderately balanced), index 4 (balanced) and index 5 (highly balanced).

$$\text{Percentage of asymmetry} = 2x\frac{\sum left - \sum right}{\sum left + \sum right}x100\% \qquad (4)$$

Table 1   Data sample per index

Index	Samples	3D Model
Index 3	9	72
Index 4	37	296
Index 5	5	40

Table 2   Brain balancing index with range of balance score

Balanced Group/Index	Percentage Difference Between Left and Right	Subjects
Moderately Balanced - 3	40.0%-59.9%	9
Balanced - 4	20.0%-39.9%	37
Highly Balanced - 5	0.0%-19.9%	5

3.2 Normality Test Using SPSS

Significant level, p which is the confidence interval for mean is 95%. Table 3 shows Shapiro-Wilk test for checking normality of the dependent variables which is max power spectral density (PSD) data for each sub bands left and right.

It shows that $p < 0.05$ for certain data in bands, so that the data distributed not in normal pattern (blue color). In the other hand, the delta right, theta right, alpha (left and right) side and beta (left and right) side of the brain fulfill the hypothesis. Some data can be seen that $p > 0.05$ and this is true for delta left side and theta left side. The data is normally distributed (red color). Therefore the result showed that mixing between normal distribution and not normal distribution, resulted to nonparametric types of data.

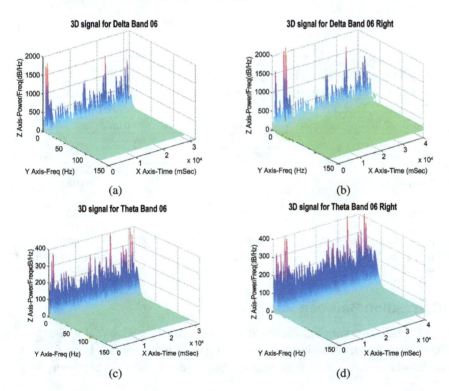

(a)   (b)

(c)   (d)

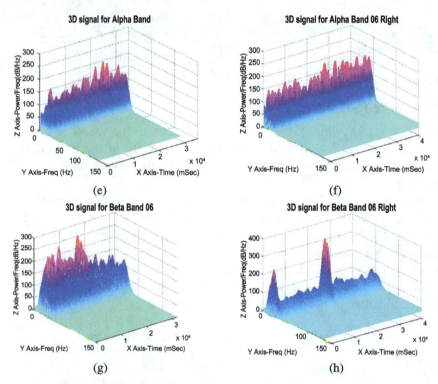

Figure 8   3D EEG models for (a) Delta (LHS), (b) Delta (RHS), (c) Theta (LHS), (d) Theta (RHS), (e) Alpha (LHS) (f) Alpha (RHS), (g) Beta (LHS) and (h) Beta (RHS)

Table 3   Shapiro-wilk test of max psd for each sub band

Sub band	Shapiro-Wilk	
	Statistic	Sig.
Delta Left	0.956	0.054
Delta Right	0.954	0.047
Theta Left	0.966	0.152
Theta Right	0.950	0.030
Alpa Left	0.946	0.022
Alpa Right	0.910	0.001
Beta Left	0.855	0.000
Beta Right	0.884	0.000

3.3 Correlation Between Sub Band

The confidence interval (significant level, p) for mean is 95%. Table 4 depict the Pearson Correlation to analyze the correlation between sub band for left and right brainwave. There was a strong positive relationship between right

Table 4   Pearson correlation value of max psd for each sub band

Sub Band	Pearson Correlation		
	Index3	Index4	Index5
Delta Left	0.816	0.526	0.613
Delta Right			
Theta Left	0.889	0.622	0.651
Theta Right			
Alpa Left	0.960	0.541	0.834
Alpa Right			
Beta Left	0.893	0.601	0.946
Beta Right			

and left side of brain for all sub bands with r > 0.5 for all sub bands at left and right side. For Index 3, alpa band is the highest correlation values (r=0.960), Index 4 theta band is the highest (r=0.622) and for Index 5 beta band is the highest value (r=0.946).

Table 3 and 4 produce results which caused by the outliers and it needs to be analyzed in the future.

4 Conclusion

In this paper, 3D EEG model is generated using signal processing and image processing. The artifact removal and band pass filter are implemented for preprocessing signal stage. The resultant images which are two-dimensional (2D) EEG image or spectrogram were constructed via Short Time Fourier Transform (STFT). Optimization, color conversion, gradient and mesh algorithms are image processing techniques have been implemented to produce this model. Results indicate that the proposed maximum PSD from 3D EEG model were able to distinguish the different levels of brain balancing indexes. The statistical analysis for LHS and RHS shows that the data is significant for all bands except delta and theta LHS. All bands from the left and right side of the brain are positively correlated. Further analysis consisting of other feature extraction technique will be done as future work.

References

[1] Y. M. Randall and C. O'Reilly, Computational Exploration in Cognitive Neuroscience: Understanding the Mind by Simulating the Brain, MIT Press London, 2000.
[2] D. Cohen, The Secret Language of the Mind, Duncan Baird Publishers, London, 1996.

[3] M. Teplan, "Fundamentals of EEG Measurement.", Measurement Science Review, vol. 2, pp. 1–11, 2002.

[4] E. R. Kandel, J. H. Schwartz, T. M. Jessell, Principles of Neural Science, Fourth Edition, McGraw-Hill, 2000.

[5] E. Hoffmann, "Brain Training Against Stress: Theory, Methods and Results from an Outcome Study", version 4.2, October 2005.

[6] R. W. Sperry, "Left -Brain, Right Brain," in Saturday Review:speech upon receiving the twenty-ninth annual Passano Foundation Award, 1975, pp. 30–33.

[7] R. W. Sperry, "Some Effects of Disconnecting The Cerebral Hemispheres," in Division of Biology, California Institute of Technology, Pasadena. California, 1981, pp. 1–9.

[8] Zunairah Haji Murat, Mohd Nasir Taib, Sahrim Lias, Ros Shilawani S. Abdul Kadir, Norizam Sulaiman, and Mahfuzah Mustafa. "Establishing the fundamental of brainwave balancing index (BBI) using EEG," presented at the 2nd Int. Conf. on Computional Intelligence, Communication Systems and Networks (CICSyN2010), Liverpool, United Kingdom, 2010.

[9] P. J. Sorgi, The 7 Systems of Balance: A Natural Prescription.

[10] R. W. Sperry, "Some Effects of Disconnecting the Cerebral Hemispheres," in *Division of Biology California Institute of Technology, Pasadena*. California, 1981, pp. 1–9.

[11] P. J. Sorgi, *The 7 Systems of Balance: A Natural Prescription* for Healthy Living in a Hectic World Health Communications Incorporated, 2002.

[12] E. R. Braverman, *The Edge Effect: Archive Total Health and Longevity*: Sterling Publishing Company, Inc., 2004.

[13] Z. Liu, L. Ding, "Integration of EEG/MEG with MRI and fMRI in Functional Neuroimaging," *IEEE Eng Med Biological Magazine*, vol. 25, pp. 46–53, 2006.

[14] U. Will and E. Berg, "Brain Wave Synchronization and Entrainment to Periodic Acoustic Stimuli," *Neuroscience Letters*, vol. 424, pp. 55–60, 2007.

[15] B.-S. Shim, S.-W. Lee, "Implementation of a 3 –Dimensional Game for Developing Balanced Brainwave," presented at 5th International Conference on Software Engineering Research, Management & Applications, 2007.

[16] Rosihan M. Ali and Liew Kee Kor, "Association Between Brain Hemisphericity, Learning Styles and Confidence in Using Graphics

Calculator for Mathematics", Eurasia Journal of Mathematics, Science and Technology Education,vol. 3(2), pp127–131, 2007.

[17] M. Hutchison, *Mega Brain Power: Transform Your Life with MindMachines and Brain Nutrients*: Hyperion, 1994.

[18] Zunairah Hj. Murat, Mohd Nasir Taib, Sahrim Lias , Ros Shilawani S. Abdul Kadir, Norizam Sulaiman and Zodie Mohd Hanafiah, "Development of Brainwave Balancing Index Using EEG", 2011 Third International Conference on Computational Intelligence, Communication Systems and Networks, pp.374–378, 2011

[19] Jansen BH, Cheng W-K. "Structural EEG analysis: an explorative study.", Int J Biomed Comput 1988; 23: 221–37.

[20] L. Sornmo, and P. Laguna, Bioelectrical Signal Processing in Cardiac and Neurological Applications. Burlington, MA: Elsevier Academic Press, 2005.

[21] N. Hosaka, J. Tanaka, A. Koyama, K. Magatani, "The EEG measurement technique under exercising", Proceedings of the 28th IEEE EMBS Annual International Conference, New York City, USA, Sept 2006, pp. 1307–1310.

[22] A. Delorme, and S. Makeig, "The EEGLAB," Internet http://www. sccn.ucsd. edu/eeglab, vol. 2, no. 004, pp. 1.2.

[23] C. Babiloni, G. Binetti, E. Cassetta, D. Cerboneschi, G. D. Forno, C. D. Percio, F. Ferreri, R. Ferri, B. Lanuzza, C. Miniussi, D. V. Moretti, F. Nobili, R. D. Pascual-Marqui, G. Rodriguez, G. L. Romani, S. Salinari, F. Tecchio, P. Vitali,O. Zanetti, F. Zappasodi, P. M. Rossin., "Mapping distributed sources of cortical rhythms in mild Alzheirmer's disease. A multicentric EEGstudy," NeuroImage, vol. 22, pp. 57–67, 2004.

[24] K. N. Diaye, R. Ragot, L. Garnero, V. Pouthas , "What is common to brain activity evoked by the perception of visual and auditory filled durations? A study with MEG and EEG co-recordings," Cognitive Brain Research,vol. 21, pp. pp. 250–268, 2004.

[25] C. Babiloni, R. Ferri, G. Binetti, F. Vecchio, G. B. Frisoni, B. Lanuzza, C. Miniussi, F. Nobili, G. Rodriguez, F. Rundo, A. Cassarino, F. Infarinato, E. Cassetta, S. Salinari, F. Eusebi, and P. M. Rossini, "Directionality of EEG synchronization in Alzheimer's disease subjects," Neurobiology of Aging, vol. 30, pp. 93–102, 2009.

[26] A. Piryatinska, G. Terdik, W. A. Woyczynski, K. A. Loparo, M. S. Scher, and A. Zlotnik, "Automated detection of neonate EEG sleep stages," Computer Methods and Programs in Biomedicine, vol. In Press, Corrected Proof.

[27] M. T. Pourazad, Z. K. Mousavi, and G. Thomas, "Heart sound cancellation from lung sound recordings using adaptive threshold and 2D interpolation in time-frequency domain," in Proceedings of the 25th Annual International Conference of the IEEE, 2003, pp. 2586–2589.

[28] Ohbuchi. R, "Incremental 3D ultrasound imaging from a 2D scanner," Conference in Biomedical Computing, Atlanta, 1990.

[29] A. I. Kochaev · R. A. Brazhe,"Mathematical modeling of elastic wave propagation in crystals: 3D-wave surfaces," Department of Physics, Ulyanovsk State Technical University, Rusia , 2011

[30] Dongmei Hao, Hongwei Zhang, and Naigong Yu "High Resolution Time-Frequency Analysis for Event-Related Electroencephalogram," Proceedings of the 6th World Congress on Intelligent Control and Automation, June 21 - 23, Dalian, China, 2006

[31] A. J. B. Tadeu*, L. Godinho, P. Santos,"Performance of the BEM solution in 3D acoustic wave scattering,"University of Coimbra, Portugal,Advances in Engineering Software vol.32 pp.629–639, 2001

Biographies

N. Fuad (Norfaiza Fuad) has received BSc Hon's in Computer Engineering from Universiti Teknologi Malaysia in 2003 and M.Sc in Computer System Engineering from Universiti Putra Malaysia in 2006. Currently is pursing PhD in Faculty of Electrical Engineering, Universiti Teknologi MARA, Shah Alam, Malaysia. Her Professional Memberships are member IEM (graduated) and IEEE. Her current research interests are in advanced signal processing with applications in biomedical, Image Processing, Biomedical, Embedded System, Microprocessor and Microcontroller and Data Encryption.

Multilingual Rules for Spam Detection

Minh Tuan Vu[1], Quang Anh Tran[1], Frank Jiang[2] and Van Quan Tran[1]

[1]*Faculty of Information Technology, Hanoi University,*
Hanoi, Vietnam
[2]*School of Engineering and IT, University of New South Wales,*
Canberra, Australia

Received 15 April 2013; Accepted 23 May 2014
Publication 4 August 2014

Abstract

In this paper, we introduced a statistical rule-based method to create rules for SpamAssassin to detect spams in different languages. The theoretical framework of generating and maintaining multilingual rules were also illustrated. The experiments were conducted against the dataset of three languages including Chinese, Vietnamese and English. The detecting achievement of multilingual rule was 89.5% for the true detection and only 3.8% for the failed alarm at the threshold of 2 while the true detection rate of single language rule was not over 61% and the failed alarm rate was up to 4.9%.

Keywords: Spam detection, multilingual rules, SpamAssassin, spam, ham.

1 Introduction

In recent years, the battle against spam e-mail is extremely fierce. Despite the anti-spam technology development, spammers keep working hard to find new strategies which help deliver unwanted messages to email users all around the world. One of these tricks is sending spam emails in different languages beside users' vernaculars. According to Message Labs' July 2009 Intelligence Report [1], in France, Netherland and Germany, spammers used spam translation technique to generate spam at 53%, 25% and 46% respectively. In China and Japan, the rates of non-English spam were even up to 63.3% and 54.7%.

Journal of Machine to Machine Communications, Vol. 1 , 107–122.
doi: 10.13052/jmmc2246-137X.122

The trick has worked relatively well because spammers dig deep into the flaw of current spam-filtering machines detecting spam based on the wordlist which is not good at dealing with multilingual emails. The report [1] also explained that by making full use of auto-translation tools, spammers have created different language spam and causes a 13% rise in overall spam in mentioned countries above.

The development of automated translation tools is natural and necessary. In order to solve this problem, the multilingual rules for spam-filtering machines should be proposed. In a recent paper, Quang-Anh Tran et al. [2] introduced a method to create Chinese rules for SpamAssassin. Although this set of rules has done a good job and been shared by thousands of email servers all around the world, it could detect the spam email in Chinese only. To surmount this ruse, we level up the method and make it multilingual. In other words, we created a system that could generate the set of anti-spam rules for different languages. The experiments were conducted with the same dataset for every language (Chinese, Vietnamese and English) and mixed type of these three ones.

The paper is structured as follows: In section II, we reviewed some approaches to filer spams in specific languages and related knowledge. Section III follows with the theoretical framework of our method. Next, the experiments are conducted and the results are compared in section IV. Finally, section V concludes the paper and discusses the future of work.

2 Related Works

2.1 SpamAssassin Rules

SpamAssassin is one of the most popular for deciding how likely an email message is spam. It filters spam based on content-matching rules. Each rule has its own score. If an email message gains enough scores (over the pre-defined threshold), it will be marked as spam.

Here is the sample of a SpamAssassin rule:

Figure 1 is an example of complete rule definition. The rule named FROM_START_WITH_NUMS checks to see if an email's FROM header starts with at least two numbers against the regular expression. It adds a score to the email's spam score if the email matches the rule. An anatomy of a rule was described in details by Schwartz (2004) [15]. In order to catch the spam effectively in specific languages, the rules should be generated based on the characteristic of those languages. That is the reason why we are aiming to build a multilingual rule set for an international environment.

```
header FROM_STARTS_WITH_NUMS      From =~ /^\d\d/
describe FROM_STARTS_WITH_NUMS    From: starts with nums

score FROM_STARTS_WITH_NUMS       0.390 1.574 1.044 0.579
```

Figure 1   SpamAssassin rule sample

2.2 Researches on SpamAssassin Rules for Specific Language

Quang-Anh Tran and his partners [2] explained in their paper that spam detections fall into two categories: rule-based and statistical-based. The first one refers to the detection performed by searching for the spam-liked pattern in the email. SpamAssissin is known as the most popular representative of ruled-based spam detection machines. The latter, on the other hand, manages to deal with a two-class categorization problem; the dataset of spam and ham is used to train the detector. Bayesian algorithms are most widely used statistical-based method for detecting spam. Androutsopoulos (2000) [4] and Graham (2002) [5] had typical works on this subject. Besides, other statistical-based methods are proposed such as Neural Network [6], Support Vector Machines [7] for spam detection.

However, each method (rule-based or statistical-based) has its disadvantages. The rule-based method is easy to share among servers (or users) but it is built manually. Thus, it is difficult to keep up-to-date with the quick changes of spam. Whereas, with the statistical-based method, it is easier to retrain the spam detector as long as the training dataset is up-to-date. However, it is impossible to share the knowledge of the detector. Therefore, they propose a hybrid method which is a trade-off rule-based and statistical-based to create the rules for detecting spam in Chinese. This method has all advantages of rule-based and statistical-based method: the quick-training for the detector and easy to share between servers.

Nguyen T.A et al. [8] showed an approach to detect Vietnamese spam based on language classification. They aimed to introduce a Vietnamese segmentation for using token selection for building a Vietnamese spam filter based on language classification and Bayesian combination to sufficiently support Vietnamese. The results on spam detection between their Vietnamese segmentation and space token segmentation were compared. Their spam detection rate is about 9% higher and the ham error rate is 3% lower.

Although both methods proposed in [2] and [8] achieve positive results, they only focused on a specific language. The question, here, is how these methods deal with the real circumstance that users receive emails in more than one languages every day and spammers keep sending multilingual spams to email users around the world.

3 Theoretical Framework

The Figure 2 illustrates how our multilingual rules are generated and maintained.

The email from different sources are classified and saved into the Spam & Ham database. The classification is carried out by email users and researchers. Because this is a kind of content-based approach, all we need of an email are the subject and body. After decoding the encoded content and strip the entire html tags attached with the email, we use Google API to detect the language of each email. For each language (only three languages including Chinese, Vietnamese and English are used in this paper), a suitable segmentation method will be called. The product of this step is a meaningful wordlist which are the output for next step. We reuse the algorithms in [2] for the rest of process which are discussed in the next part.

The multilingual rules set are generated automatically by three steps: Pattern retrieval, Pattern Selection and Score Assignment.

3.1 Pattern Retrieval

As we mentioned above, each language has its own way to split sentences into meaningful words. For some languages such as English, French or Germany, words can be identified easily by the space. However, with Vietnamese,

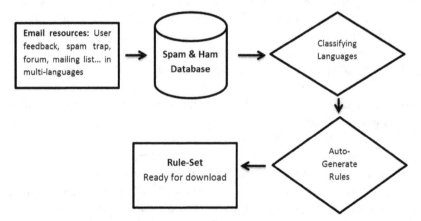

Figure 2   Process of generating multilingual rules

Chinese or some other Asian languages, it is impossible because they have special linguistic united known as syllable ("Tiếńg" in Vietnamese or "hanzi" in Chinese).

In order to achieve the highest effectiveness when processing the email content, we used Google Translate API [9] to detect the language of the email. Although the API works well and is easy to use, it is not free. Then, we only used for the experimental period. For further usage, we consider some other solutions such as Lingua::Identify available at [10] and Guess-Language [11]. In this paper, we only implemented the segmentation for three languages: Chinese, Vietnamese and English. However, we are working hard to propose a multilingual word segmentation method as Guo-Wei Lee mentions in his research [12].

With Chinese emails, we applied exactly the Chinese segmentation technique used in [2] which is based on methods: Dictionary-based, Maximum Matching; and from left to right.

With Vietnamese emails, we dealt with the word segmentation by a program proposed by Phuong Le-Hong [13]. This program works on the Vietnamese text file or folder and exports the meaningful wordlist to the XML format. It is quite straightforward to read the wordlist from this XML file.

It is much simpler to split words in English emails because words are separated by spaces. We just found and replaced the space character with the new line character and eliminated all punctuations in the sentence.

3.2 Pattern Selection

After classifying the email by languages and extracting the meaningful words, we applied some pattern selection methods to select good patterns for subject rules and body rules, individually. We could not find any difference among selecting pattern of Chinese, Vietnamese and English emails; thus, we once again reused the pattern selection algorithms in [2].

In spite of being based on the traditional pattern selection method by Yang [14], there are some changes in the approach. Only the spam-liked patterns were used to detect spam. As a result, the formula for selecting pattern was modified. The V_{ts} and V_{th} are computed as follows according to Conditional Probabilities and Bayes's Theorem:

$$V_{ts} = P(E \mid H) = \frac{P(E \wedge H)}{P(H)} \tag{1}$$

$$V_{th} = P\left(\overline{E} \mid H\right) = \frac{P\left(\overline{E} \wedge H\right)}{P\left(\overline{H}\right)} \tag{2}$$

In which:

- V_{ts} and V_{th} can best evaluate the connection between pattern t and spam, pattern t and ham, namely.
- Top N pattern that have highest value of ratio $Rt = V_{ts}/V_{th}$ are chosen.
- N is the size of the rule set, which is a factor that control the performance of the rule set.
- E is a hypothesis that a message occurs as spam.
- H is a hypothesis that a message occurs as ham.

Given a spam and ham datasets, for a pattern t, A and B are the number of times that spam and ham messages contain t, respectively; C and D are the numbers of times spam and ham messages do not contain t, respectively. The values of the probabilities in (1) and (2) are computed as follows.

$$P\left(E\right) = \frac{A + C}{A + B + C + D} \tag{3}$$

$$P\left(\overline{E}\right) = \frac{B + D}{A + B + C + D} \tag{4}$$

$$P\left(H\right) = \frac{A + B}{A + B + C + D} \tag{5}$$

$$P\left(E \wedge H\right) = \frac{A}{A + B + C + D} \tag{6}$$

$$P\left(\overline{E} \wedge H\right) = \frac{B}{A + B + C + D} \tag{7}$$

3.3 Score Assignment

The rules are created on the basis of the selected set of spam-liked patterns. There are two types of rules: Body rule and Subject rule. The Fast SpamAssassin Score Learning Tool by Henry Stern [15] is used to assign the score to each rule.

According to the illustration of Quang-Anh Tran et al [2], The "Stochastic Gradient Descent" method of training a neutral network was implemented. The program uses a single perceptron and (8) a logsig activation function (9) to map the weights to SpamAssassin score space.

$$f\left(x\right) = \int_{i=1}^{N} w_i x_i \tag{8}$$

$$y\left(x\right) = \frac{1}{1 + e^{-f(x)}} \tag{9}$$

where w_i represents the score for rule i and x_i describes whether a given message activates rule i or not, the transfer function (8) returns the message's score. The gradient descent is employed to train the neural network. The parameter of the network is tuned iteratively to ensure that the rate of mean error always decreases. Without getting into calculus, the error gradient for a perceptron with a linear transfer function, logsig activation function and mean squared error function is as follows:

$$E\left(x\right) = y\left(x\right)\left(1 - y\left(x\right)\right)\left(y_{exp} - y\left(x\right)\right) \tag{10}$$

And the weights are updated using the function:

$$w_i = w_i + \alpha E\left(x\right)x_i \tag{11}$$

In which, α is a learning rate. The implementation uses the so-called "Stochastic gradient descent" method which does incremental updates by walking through the training set randomly rather than doing one batch update per epoch because the SpamAssassin rule hits are spares.

4 Experiments

4.1 Dataset

The data we used to conduct the experiments is divided into 4 groups.

E-mails come from email users' personal inboxes and are classified as spam and ham manually by authors. We store all emails in the MySQL database. The spam and ham in each language are saved in separated table with the same structure (ID (PK), Subject, Body, Status, Date), then, there are eight tables serving the experiments.

Firstly, the experiments are conducted with three first groups (Group 1, 2 and 3) to create the rule for corresponding language. The rule set is tested based on single language dataset only. The results are saved for the

Table 1 Dataset description

Group	Num. of Spams	Num. of Hams	Language
1	200	200	Chinese
2	231	251	Vietnamese
3	274	202	English
4	705	653	Multi-languages

comparison (1). Next, the single language rule sets are tested based on data group 4 (multi-language emails) to evaluate the efficiency (2). Finally, the mixed languages rule set is generated based on data group 4. The effectiveness of this rule set is recorded and compared with the result (1) and (2).

4.2 Single-language Rule Set Creation

The procedures of generating rule set for specific are applied exactly mentioned in section 3.

For Chinese rules, the experiment is based on the data group 1 with 200 hams and 200 spams. The spam detection rate (Spam Recall) and the failed alarm rate (Ham Error) are illustrated in Table 2.

The Chinese rule gives the best result with the threshold equal to 2.5 at which the positive true rate is 91.5% and the failed alarm rate is eliminated.

The experiment conducted based on the Vietnamese (generating rules and testing rules totally with Vietnamese dataset – group 2) also brings positive results.

Table 2   Performance of Chinese rule with Chinese dataset

Threshold	Spam Recall	Ham Error
0.5	93.5%	30.5%
1	91.5%	9.0%
1.5	91.5%	5.5%
2	91.5%	4.0%
2.5	91.5%	0.0%
3	91.5%	0.0%
3.5	85.0%	0.0%
4	75.5%	0.0%
4.5	71.0%	0.0%

Table 3   Performance of Vietnamese rule with Vietnamese dataset

Threshold	Spam Recall	Ham Error
0.5	90.5%	34.7%
1	87.4%	27.9%
1.5	83.1%	11.2%
2	81.4%	2.4%
2.5	81.4%	0.0%
3	78.4%	0.0%
3.5	73.6%	0.0%
4	66.2%	0.0%
4.5	59.3%	0.0%

Table 4   Performance of English rule with English dataset

Threshold	Spam Recall	Ham Error
0.5	98.5%	81.2%
1	97.1%	50.5%
1.5	96.0%	24.3%
2	95.6%	5.0%
2.5	95.3%	0.0%
3	93.1%	0.0%
3.5	87.6%	0.0%
4	82.8%	0.0%
4.5	60.2%	0.0%

At threshold 0.5, the spam recall rate is really high (90.5%) but the ham error rate is unacceptable (up to 34.7%). However, when we increase the threshold, the result is better and better. Especially, the ham error rate falls significantly at the threshold 2 (2.4%) while the spam recall stay unchanged in comparing to the previous threshold (81.4%).

We did the same thing to generate the English rule set and then recorded the result after testing the rule based on English emails only. The results are displayed in the Table 4.

The English rule set works extremely effectively at the threshold of 2.5. At this point, the positive true rate stays high over 95% while the ham error is totally eliminated.

On finishing the experiment to generate the SpamAssassin rule and to test these rule set with the corresponding language, we gained really positive results on true spam detection rate and failed alarm rate. However, whether these rule sets still work well with multi-language dataset? The answer is coming with the next experiment.

4.3 Single-language Rule Set Tested with Multi-language Emails

Three sets of rule in Chinese, Vietnamese and English are tested with the data group 4 which contains 705 spams and 653 hams in multi-languages in order to evaluate the efficiency in a multilingual environment. Table 5 shows the result of how Chinese, Vietnamese and English rule sets works.

The statistic shows obviously that when working with the multilingual email dataset, all sets of rules give very poor performance in the true positive rate, especially, the Chinese rule which detects only 24.5% at threshold 0.5 in comparing to over 93% of the last experiment. English rule and Vietnamese are better at spam detecting but the result is far lower than those when working

Table 5   Performance of single language rule with multilingual dataset

Threshold	Chinese		Vietnamese		English	
	Spam Recall	Ham Error	Spam Recall	Ham Error	Spam Recall	Ham Error
0.5	24.5%	0.2%	60.6%	3.5%	51.9%	4.9%
1	21.8%	0.2%	59.0%	2.0%	49.9%	1.8%
1.5	20.7%	0.0%	54.8%	0.2%	44.3%	1.1%
2	19.0%	0.0%	53.6%	0.2%	42.4%	0.3%
2.5	18.2%	0.0%	49.4%	0.0%	41.6%	0.0%
3	16.9%	0.0%	46.8%	0.0%	40.7%	0.0%
3.5	16.7%	0.0%	38.4%	0.0%	39.9%	0.0%
4	16.0%	0.0%	20.7%	0.0%	32.5%	0.0%
4.5	15.6%	0.0%	14.8%	0.0%	22.1%	0.0%

with single language dataset. The explanation for the fall is quite clear and straightforward. The rule generated from specific language can detect the spam effectively in that language only. Therefore, we are expecting a multilingual rule set that can detect spam effectively among ton of multi-language emails.

4.4 Multilingual Rule Set

The final experiment is to generate and to test the rule set from the data group 4 which contains the email in three languages Chinese, Vietnamese and English.

After classifying the language of each email, we did the word segmentation for the set of emails in the same language. The pattern selection chooses the best pattern for evaluating the whether a pattern is spam-liked or not. Based on the selected pattern the rule set is generated automatically. The Fast SpamAssassin Score Learning Tool will handle the rest by assigning the score for each rule. Applying these steps on the multi-language dataset, we gained a set of multilingual rule for detecting the spam.

Table 6 illustrates the result of the test detecting spam based on the multilingual rule set. At the first level of the threshold, although the spam recall rate is highest (94.17%), the ham error rate is up to 48.80%. It is intolerant for a set of SpamAssassin rules. However, the result is much better when the threshold increase to 2.5. The true positive rate is 89.40% and the false positive rate is 0%. At this threshold, with the same multi-language dataset, the performance of Chinese rule, Vietnamese rule and English rule are 18.2%, 49.4% and 41.6% namely. This comparison proves that the rule generated from the multilingual dataset works much more effectively than the one generated based on the single language only.

Table 6 Performance of multilingual rule with multilingual dataset

Threshold	Spam Recall	Ham Error
0.5	94.17%	48.80%
1	92.00%	29.13%
1.5	90.20%	13.67%
2	89.50%	3.80%
2.5	89.40%	0.00%
3	87.67%	0.00%
3.5	82.07%	0.00%
4	74.83%	0.00%
4.5	63.50%	0.00%

5 Remarks

With a number of experiments carried out above, three sets of single-language rules are generated. The results in the Tables 3, 4 and 5 show that these sets of rules work effectively when dealing with the set of single-language dataset (Most of emails are in only one language). At the same threshold 2.5, the Chinese rules can detect up to 91.5% spam, the Vietnamese rules detect 81.4% spam and the result of English rules are 95.3% while the failed alarm is 0%.

However, when applying these sets of single-language rules in detecting multilingual spam, the results are not good at all. In detail, at the threshold 2.5, the percentage of spam detecting of Chinese, Vietnamese and English rules are 18.2%, 49.4% and 41.6%. The reason for this drop is clear. Each set of single-language rules is generated based on the corresponding language dataset. It means the rule can deal with spam in that language only. Therefore, a set of multilingual rule is considered and evaluated. This set of rules is built from the multilingual dataset including Chinese, Vietnamese and English.

An experiment is run to evaluate the performance of multilingual set of rules. The result is positive and promising. At threshold of 2.5, the rate of spam detecting is 89.40% while the ham error is eliminated. From these findings, it is shown that that effectiveness of a multilingual set of rule in detecting spams when applying in an international working environment.

6 Conclusion

Generating the rule for spam detection based on a specific language is a proper approach to fight against spammers. However, in order to deal with an email server receiving emails in more than one language, we need an extended solution. Therefore, we upgraded the method proposed in [2] to implement the system that is able to generate automatically the multilingual rule based on the multi-language dataset. The experiment results show that these rules help SpamAssassin detect spam more exactly in comparison with the ones generated based on single language dataset.

Despite of the positive results achieved, there are some issues we need to deal with in the future. Firstly, a new method to detect the language of the email should be analysed. The cost for the current one is so high. Secondly, we are expecting a better algorithm to retrieve the pattern of raw emails. Finally, it would be a big problem for the word segmentation if the system faces up to a large number of languages due to the lack of a common segmentation method as mentioned in [12].

Acknowledgments

This research was supported by the Vietnam National Foundation for Science and Technology Development (NAFOSTED) under project number 102.01-2010.09. This work received tremendous support from the Vice-chancellors' research initiatives from the University of New South Wales (UNSW).

References

[1] Multi-language Spam-A New Trend Among Spammers. Available: http://www.spamfighter.com/News-12908-Multi-language-Spam-A-New-Trend-Among-Spammers.htm

[2] Tran, Q. A., Duan, H. X. Li, X., 'Real-time statistical rules for spam detection' IJCSNS International Journal of Computer Science and Network Security, VOL.6 No.2B, pp 178–184, February 2006.

[3] Androutsopoulos I., Koutsias, J., Chandrinos, K.V., Paliouras, G., Spyropoulos, C.D., 'An evaluation of Naive Bayesian anti-Spam filtering', Proceedings of the Workshop on Machine Learning in the New Information Age, pp 9–17, 11th European Conference on Machine Learning, Barcelona, Spain, 2000.

[4] Graham, P., 'A plan for spam. Web document' (2002). Available: http://www.paulgraham.com/spam.html.

[5] Drucker, H., Wu, D., Vapnik V., 'Support Vector Machines for spam categorization', IEEE Transaction on Neural Networks.10(5), 1048–1054, 1999.

[6] Özgür, L., Güngör, T., Gürgen, F., 'Adaptive anti-spam filtering for agglutinative languages: a special case for Turkish', Pattern Recognition Letters, 1819–1831 25. 2004.

[7] Nguyen T.A., Tran Q.A., Nguyen N.B., 'Vietnamese spam detection based on language classification', HUT-ICCE 2008 - 2nd International Conference on Communications and Electronics, Hoi An, Vietnam, 2008.

[8] Google Translate API. Available: https://developers.google.com/translate/

[9] Lingua::Identify. Available: http://search.cpan.org/ambs/Lingua-Identify-0.51/lib/Lingua/Identify.pm

[10] Guess-Language. Available at http://code.google.com/p/guess-language/

[11] Guo-Wei Lee, 'A Mechanism for Filtering Multilingual Spam Mail based on Decision Tree and Integrated Feature Selection Algorithm', 1997.

[12] Phuong Le-Hong, et al, 'A hybrid approach to word segmentation of Vietnamese texts', Proceedings of the 2nd International Conference on Language and Automata Theory and Applications, LATA , Springer LNCS 5196, Tarragona, Spain, 2008.

[13] Yang, Y., Pedersen, J.O., 'A comparative study on feature selection in text categorization', Proceedings of the 14th International Conference on Machine Learning, pp 412–420, 1997.

[14] The Fast SpamAssassin Score Learning Tool. Available: http://spamassassin.apache.org

[15] Schwartz, 'SpamAssassin', 2004, O'Reilly.

Biographies

Vu Minh Tuan is a lecturer and a research coordinator of the Faculty of Information Technology at Hanoi University. Currently, he also is working as the project manager at the software department of IP Communications, JSC.

In 2012 he started his Master of Science from University of Central Lancashire, UK. He was a student of Hanoi University from 2006 to 2010. He has been working on Vietnamese rules for SpamAssassin and some projects in the field of AntiSpam, email system and data mining.

Tran Quang Anh is an associate professor of information technology from Hanoi University. He obtained Ph.D and Master degrees at Tsinghua University in 2003 and 2001 respectively. He finished his bachelor at Huazhong University of Science and Technology in 1997. His research interests include network security evolutionary algorithms and field-programmable gates array.

Dr. Jiang received his B.Sc. degree in System & Control Engineering and M.Sc. degree (by research) in Computer Science Engineering on 1997 and 1999 respectively in China and Australia. With a success of holding an Australian Postgraduate Award (APA) scholarship, he completed his PhD degree in communication engineering and software engineering at University of Technology, Sydney (UTS) in 2008. Prior to joining into UTS, Dr. Jiang has 5 years' hardware and software working experience in the VoIP industry in Sydney, Australia from 1999 to 2003. After his PhD, he was employed as a research assistant and later a research fellow for overall 3 years in Faculty of Engineering and IT (FEIT), UTS. Additionally, he has 5 years of teaching experiences as a lecturer and 2 years of subject coordinator in UTS. Currently, he works in the University of New South Wales (UNSW) as a lecturer and a full-time UNSW Vice-Chancellor appointed Research Fellow. He has published over 60 international journal and conference papers in the field of computational intelligence and its applications. His current research interests include data analytics, bio-inspired algorithms and metaheursitics, Underwater communication, Network Security, Autonomic communication networks, Intelligent and mobile agents, Network Protocols.

Van Quan, Tran Quan graduated from Hanoi University (HANU) on 2012 with the B.Sc. degree in Information technology and currently (2014) is an Information System Design master student at University of Central Lancashire, UK. He has teaching experience as a teacher assistant at HANU and more than 2 years' experience working as a developer. His current research interests include: text mining, voice recognition and human-computer interaction.

Maximisation of Correct Handover Probability and Data Throughput in Vehicular Networks

L. Banda and M. Mzyece

Department of Electrical Engineering and The French South African Institute of Technology (F'SATI), Tshwane University of Technology, Gauteng, South Africa

Received 15 April 2014; Accepted 24 May 2014
Publication 4 August 2014

Abstract

In the past decade, the networking and automobile industry has experienced the emergence of vehicular networks which were developed under the Intelligent Transportation Systems (ITS) to provide a plethora of safety and non-safety related applications. The provision of seamless mobility and session continuity is one of the major challenges for the transmission of ITS applications in vehicular networks. This is more critical when a communicating node moves from one subnet to another, a process referred to as inter-subnet handover. In such a case, we deal with the problem of fast and seamless handover support for better Quality of Service (QoS) provisioning especially, for throughput-sensitive and delay-intolerant ITS applications. In this paper, we propose a scheme aimed at improving the handover performance in IP-based vehicular networks by maximising the correct handover probability ($P_{Correct}$) and increasing the data throughput during inter-subnet handovers. We demonstrate the impact of mobile node's speed and direction of motion on both $P_{Correct}$ and data throughput via numerical analysis.

Keywords: Handover, Internet Protocol (IP), Correct Handover Probability, Data Throughput, Vehicular Networks.

Journal of Machine to Machine Communications, Vol. 1 , 123–144.
doi: 10.13052/jmmc2246-137X.123

1 Introduction

Wireless vehicular communication networks were developed under the Intelligent Transportation System (ITS) technology to improve the safety, efficiency and environmental sustainability of transportation systems. ITS applications can be classified into two broad categories: safety and non-safety applications. Safety applications are concerned with sharing of information on accidents, weather forecast, traffic congestion, road works and other precautionary measures within and among communicating vehicles. On the other hand, non-safety applications involve infotainment services like on-board Internet access, instant messaging, remote access of servers, electronic toll payments and so forth [1, 2].

Vehicular networks are an emerging ITS technology integrating wireless communication into the automobile industry. As a result, different standardisation bodies (e.g., IEEE and IETF) have been working in collaboration with various consortia (e.g., Car-to-Car Communications Consortium (C2C-CC [3]) on several issues concerning vehicular communications networks. Generally, vehicular networks consist of two types of wireless communication nodes which are Dedicated Short Range Communication Devices (DSRC). These are: On-Board Units (OBUs) mounted on vehicles and interlinked with Application Units (AUs) such as laptops, tablets and smart phones used by passengers; and Road Side Units (RSUs) found on fixed network infrastructure such as cellular Base Stations (BSs) or WiFi Physical Hot Spots (PHSs). The main modes of wireless communication present in vehicular networks are: vehicle-to-vehicle (V2V), vehicle-to-roadside (V2R) and inter-roadside (R2R) communication. Figure 1 shows a typical vehicular network architecture.

In wireless communication networks, handover management and location management are the two basic functionalities performed by the mobility management protocols for seamless mobility to be achieved. Handover management means keeping communication between two nodes alive as the Mobile Node (MN) moves freely and changes its point of attachment to the network. On the other hand, location management involves identifying the current location of an MN and keeping track of its location changes as it moves within the network [4]. Mobility management is vital in IP-based wireless networks, particularly during inter-subnet handovers when an MN changes its point of attachment to the Internet. For IP session continuity to be maintained during inter-subnet handovers, the IP-layer mobility management protocols should be able to support the rapid wireless link changes and the

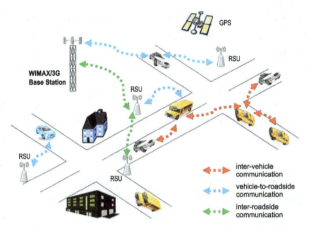

Figure 1   Typical vehicular network architecture [3]

fast IP configuration procedures [2, 5]. However, most current IP mobility management schemes fall short in that respect and as a consequence, a degradation in the Quality of Service (QoS) is registered for most applications.

The movement pattern of vehicles plays a critical role in the modelling and performance analysis of wireless IP networks. This study involves vehicle-to-roadside infrastructure (i.e. V2I) communication and focuses on network layer inter-subnet handovers when a single MN changes its point of attachment to the global Internet. The large signalling delays associated with inter-subnet handovers can be detrimental to throughput-sensitive and delay-intolerant ITS applications [6, 7]. In this paper, we propose a mobility model aimed at reducing the signalling delay thereby, maximising the probability of correct handover initiation and increasing the data throughput during inter-subnet handovers in wireless vehicular scenarios.

The remainder of the paper is organised as follows. Section 2 presents the background and the related work found in the literature. Section 3 provides a system description and modelling of the proposed scheme. Section 4 presents the simulations methodology and numerical analysis. Section 5 finally, concludes the paper.

2 Background and Related Work

Mobility models are used to represent the movement patterns of mobile users in a network. They are employed to document how the MN's location, velocity and direction of motion change with time. Various approaches can be adopted in modelling the movement of vehicles, and they all undergo a common trade-off between precision and complexity. The research community in both industry and academia have devoted considerable efforts in their quest to address the problem of severe signalling overheads and inaccurate mobility modelling which lead to miss-allocation of the limited radio resources during inter-subnet handovers. To this end, numerous standard and non-standard mobility models have been outlined in the literature [8].

2.1 Standard Vehicular Mobility Models

From the analytical and mathematical point of view, the following mobility models have been standardised for the wireless vehicular environment.

1. *Constant Speed Motion (CSM) Model*

This model describes a random vehicular movement on a graph, representing a road topology. No particular constraint is forced on the graph nature so as to display different levels of realism [9]. The vehicle's motion is structured in movements between vertices of the graph referred to as destinations which are randomly selected. At the start of each trip, a vehicle i chooses its next destination, computes the route to it by running a shortest path algorithm on the graph with link costs possibly influenced by parameters such as road length, speed limits, traffic congestion and so forth. The vehicle sets its speed to a value given by

$$v_i = v_{min} + \eta(v_{max} - v_{min}) \tag{1}$$

where v_{min} is the minimum allowed/desired speed, v_{max} is the maximum allowed/desired speed and η is a uniformly distributed random variable in the range [0, 1]. The speed v_i is selected once at the beginning of each trip and kept constant until the next destination is reached.

2. *Manhattan Model*

The Manhattan model [10] adds complexity to the speed management of the constant speed motion (CSM) model by updating the vehicles speed according to the following conditions.

$$v_i\,(t + \triangle t) = \begin{cases} v_{i+1}\,(t) - a/2, & \text{if } \triangle x_i(t) \le \triangle x_{min} \\ \tilde{v}_i\,(t + \triangle t), & \text{otherwise} \end{cases} \tag{2a}$$

$$\tilde{v}_i\,(t+\triangle t) = \min\left\{\max\left\{v_i\,(t)+\eta a\triangle t,\,v_{\min}\right\},\,v_{\max}\right\} \qquad (2b)$$

where η is the same uniform random variable which was introduced before and a is the vehicle's uniform acceleration. $\triangle x_{min}$ and $\triangle x_{max}$ are the minimum and maximum allowable inter-vehicle spacing, respectively. The Manhattan model thus, adds some acceleration-bounded randomness in the velocity update and from Equations (2) above, imposes speed limitation to prevent vehicles from overlapping [11]. The main drawback of this model is that it lacks the pause handling at road inter-sections.

3. *Freeway Model*

The Freeway model is designed for road topology graphs representing non-communicating, bi-directional, multi-lane freeways traversing the entire simulation area [12]. According to [9], the movement of each vehicle is restricted to the lane it is moving on, and the following speed management rules apply to vehicle i:

- Speed update: The speed is varied by a random acceleration of maximum magnitude a. If a random variable η is defined to be uniformly distributed in the range $[-1, 1]$, then this rule can be expressed as

$$v_i\,(t+\triangle t) = v_i\,(t)+\eta a\triangle t \qquad (3)$$

- Speed bounding: At any particular moment during motion, the speed of a vehicle cannot be lower than a minimum value ($v_{\min}$) and cannot exceed a maximum value ($v_{\max}$) . This constraint can be formulated as

$$v_i\,(t+\triangle t) = \min\left\{\max\left[v_i\,(t+\triangle t),\,v_{\min}\right],\,v_{\max}\right\} \qquad (4)$$

- Speed reduction: In order to avoid overlapping (collision situation) with the front vehicle, a minimum safety distance must be observed. This can formally be enforced as

$$v_i\,(t+\triangle t) = \begin{cases} v_{i+1}\,(t)-a/2, & \text{if } \triangle x_i(t) \le \triangle x_{\min} \\ \tilde{v}_i\,(t+\triangle t), & \text{otherwise} \end{cases} \qquad (5)$$

Each vehicle starts its movement at one end of a lane, with a speed that is initially selected as being uniformly distributed in the range $[v_{min}, v_{max}]$, and stops with it once it reaches the other extreme end of the same lane. Then a new movement is started on a randomly selected lane and the process is repeated.

2.2 Proposed Non-standard Vehicular Mobility Models

Several non-standard vehicular mobility models have been proposed in the literature. However, we will only highlight a few key examples that are directly relevant to our work.

An analysis of improving data throughput and increasing the handover probability during network layer handovers is given in [13]. The proposed method aims at improving handover performance by maximising both the handover probability and data throughput in vehicular scenarios. Our current study is an extension of the method proposed in [13].

In [7], an enhanced network layer handoff performance meant to minimise the handoff failure probability in next generation wireless systems is proposed. Based on the information of false handoff probability, the authors analyse its effects on mobile speed and handoff signalling delay.

Authors of [14] propose a scheme meant to provide high accurate prediction of the next crossing cell that the MN is going to go, in order to avoid over-reservation of the limited system resources thereby, reducing wastage of such resources.

In [15], a method of minimising handoff latency by angular displacement method using Global Positioning System (GPS) based map is proposed. This is achieved by minimising the number of access points (APs) scanned by the mobile node (MN) during each handover procedure.

A link layer assisted handover algorithm meant to enhance Received Signal Strength (RSS) value and thus, reduce the handover latency and handover failure is proposed in [16]. This algorithm employs an approach where access points used in a wireless LAN environment and a dedicated MAC bridge are jointly used to achieve packet loss without altering the Mobile IP specifications.

A new enhanced Handoff Protocol for Integrated Networks (eHPINs) which localizes the mobility management enabling fast handover is introduced in [17]. The eHPINs scheme alleviates the service disruptions during roaming in heterogeneous IP-based wireless environments thereby, improving the QoS for real-time applications in such networks.

In [18], a scheme called Simplified Fast Handover in Mobile IPv6 Networks (SFMIPv6) is proposed. This scheme is an enhancement of the FMIPv6 standard which significantly reduces the anticipation time for fast handovers thereby, increasing the probability of predictive fast handover execution.

3 Description and Modelling of the Proposed Scheme

The access network of the proposed scheme is based on next generation wireless systems with radio base stations (BSs) providing Internet connectivity to on-board users. Furthermore, the proposed scheme adopts the concept of sectorisation by dividing the coverage area of a single BS into six cell sectors using direction antennas. In addition, we assume that the vehicle under investigation is mounted with a Global Positioning System (GPS) device to keep track of the real-time mobility parameters of the vehicle such as speed, position and direction of motion [7, 13].

3.1 Handover and Non-handover Regions

To increase the rate and accuracy of candidate router discovery, each cell sector is divided into two regions, which are: the handover region (near the cell border) and the non-handover region (near the transmitter). In the non-handover region, the MN is under the full coverage of the serving BS and thus, there is no need for handover execution. However, when the MN enters the handover region, it experiences different BS signal probes with varying signal strength and quality which can trigger the handover process [19]. Figure 2 shows an example of handover and non-handover regions between for BSs belonging to two different subnets.

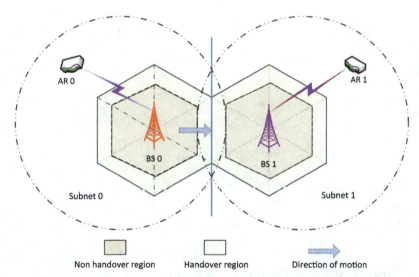

Figure 2   Handover and non-handover regions in the proposed scheme

When a vehicle moves from the non-handover region to the handover region as shown by the arrow in the Figure 2, handover preparations start as the MN scans and observes potential candidate BSs through probes and pilot signals. In the BS probes are neighbour cell information such as network ID, subnet ID, BS ID and cell ID. In addition, the attached GPS device traces the position, speed and direction of the vehicle which are recorded and stored in the memory of the Application Units (AUs). The target cell information and the GPS stored information are transmitted to the current Access Router (cAR) via the current BS (cBS). During inter-subnet handover preparation, the cAR requests the target Access Router (tAR) to reserve radio resources for the MN to use at the target BS (tBS). At the same time, a bi-directional tunnel is created between the two ARs and the cAR transfers data packets meant for the MN to the tAR.

3.2 Cell Overlap Region

The coverage area of a single BS is represented by a regular hexagon. For an inter-subnet handover process, we assume the two adjacent regular hexagons representing service areas of two neighbouring BSs to be overlapping [7, 18]. Therefore, the near-to-reality circular cells do overlap thereby, creating a cell overlap region where handover execution takes place. Figure 3 shows the cell overlap region for two radio cells belonging to two different subnets.

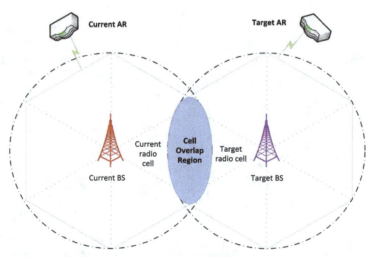

Figure 3   Cell overlap region for an inter-subnet handover scenario

3.3 Mobility Modelling

From Figure 3 above, we geometrically formulate the mobility model of the vehicle as it enters the cell overlap region during an inter-subnet handover process. This is illustrated in Figure 4 below. RSS_{Th} and RSS_{min} depict the received signal strength (RSS) threshold value required to initiate a handover and the MN's minimum RSS value needed to communicates successfully with a BS, respectively [7, 18]. The distance travelled by the MN from the time RSS_{Th} is detected to the time a handover takes place is given by d with α being the angular direction of travel. We assume the vehicle under investigation enters the cell overlap region at point B and moves in a straight line through the overlap region with an angular displacement of $\alpha\epsilon[-\theta, \theta]$ radians with respect to line BI in Figure 4.

In the figure, when the vehicle crosses line DG, an inter-subnet handover occurs. This consists of link-layer handover from current BS to target BS followed by network-layer handover from current Access Router (AR) to target AR. By geometry, $EI = AI = r$, (i.e., one side of a regular hexagon = radius of the circle circumscribing the hexagon). Let $CO = \delta$ (i.e., size of intrusion of one hexagon into another).

$$AH = r\cos 30°$$
$$= \sqrt{3}\times r/2$$

$$BC = HI = AI - AH \qquad (6)$$
$$= (2r-\sqrt{3}r)/2$$

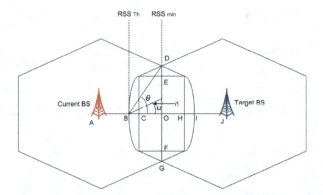

Figure 4   Handover mobility model analysis

We assume that a handover takes place only when the vehicle crosses line DG in Figure 4.

$$BO = BC + CO$$
$$= (2r - \sqrt{3}r + 2\delta)/2 \qquad (7)$$

Further, from the figure, we can deduce that

$$DE = \delta \tan 30°$$
$$= \delta/\sqrt{3}$$
$$DO = DE + EO$$
$$= \frac{\delta}{\sqrt{3}} + \frac{r}{2}$$
$$= (2\delta + \sqrt{3}r)/2\sqrt{3}$$
$$\tan \theta = \frac{DO}{BO}$$
$$= \frac{\sqrt{3}r + 2\delta}{\sqrt{3}(2r - \sqrt{3}r + 2\delta)} \qquad (8)$$

3.4 Correct Handover Initiation Probability

If we let the initial speed of the vehicle to be v_i at an initial angular displacement of θ_i, then the direction of motion is a uniformly distributed random variable with the Probability Density Function (PDF) given by

$$f_\theta(\theta) = \begin{cases} 1/2\pi \; ; & -\pi \le \theta \le \pi \\ 0 \quad ; & \text{otherwise} \end{cases} \qquad (9)$$

Handover to the target radio cell occurs only if the vehicle's direction of motion from point B in Figure 4 is in the range $(-\theta, \theta)$, where $\theta = tan^{-1}[\frac{\sqrt{3}r + 2\delta}{\sqrt{3}(2r - \sqrt{3}r + 2\delta)}]$. The probability of correct handover initiation $(P_{correct})$ when the vehicle is at point B is calculated as follows.

$$P_{correct} = \int_{-\theta}^{\theta} f_\theta(\theta) \; d\theta$$
$$= \frac{\theta}{\pi}$$

$$= tan^{-1} \left\{ \frac{\sqrt{3}r + 2\delta}{\sqrt{3}(2r - \sqrt{3}r + 2\delta)} \right\} / \pi \tag{10}$$

Therefore, the probability of correct handover initiation is dependent on the value of both δ and r. When $\delta = 0$, probability of correct handover initiation has a constant value ($P_{correct} = 0.417$). According to [7], handover failure probability increases as handover signalling delay increases. As a consequence, probability of handover success decreases as handover signalling delay increases. In Figure 4, for the direction of motion of the vehicle from B, where $\alpha \in [-\theta, \theta]$, the time taken t, from the moment RSS_{TH} is detected by the MN to the moment a handover process starts is given by

$$t = (2r - \sqrt{3}r + 2\delta)/(2v\cos\alpha) \tag{11}$$

where v is the vehicle's constant speed. The PDF of the angular direction, α is given by

$$F_\alpha(\alpha) = \begin{cases} \frac{1}{2\theta_1}; & where -\theta_1 < \alpha < \theta_1 \\ 0; & otherwise \end{cases} \tag{12}$$

From Equation (11), t is a function of α i.e., $t = y(\alpha)$ and therefore, (11) can be expressed as

$$y(\alpha) = \frac{(2r - \sqrt{3}r + 2\delta)}{2v\cos\alpha} \tag{13}$$

According to [7], we can calculate the PDF of t as

$$F_t(t) = \sum \frac{F_\alpha(\alpha_i)}{|y'(\alpha_i)|} \tag{14}$$

where α_i are the two roots of the equation $t = y(\alpha)$ in the interval $[-\theta_1, \theta_1]$. In either case, $F_\alpha(\alpha_i) = \frac{1}{2\theta_1}$, for $i = 1$ and 2. Consequently, $F_t(t)$ can be expressed as

$$F_t(t) = \frac{1}{\theta_1 |y'(\alpha_i)|} \tag{15}$$

where $y'(\alpha)$ is the first derivative of $y(\alpha)$ and is given by

$$y'(\alpha) = t \tan \alpha$$
$$= t\sqrt{(sec^2\alpha - 1)}$$

$$= t\sqrt{\left\{\left(\frac{2vt}{2r - \sqrt{3}r + 2\delta}\right)^2 - 1\right\}} \qquad (16)$$

From Equations (15) and (16), the PDF of t becomes

$$F_t\left(t\right) = \begin{cases} \dfrac{2r-\sqrt{3}r+2\delta}{\theta_1 t\sqrt{\left[(2vt)^2-\left(2r-\sqrt{3}r+2\delta\right)^2\right]}}, & \Phi_1 < \Phi_2 \\ 0, & otherwise \end{cases} \qquad (17)$$

where $\Phi_1 = \frac{2r-\sqrt{3}r+2\delta}{2v}$ and $\Phi_2 = \frac{\sqrt{\frac{(2r-\sqrt{3}r+2\delta)^2}{4}+\frac{(\sqrt{3}r+2\delta)^2}{12}}}{v}$.

According to [7, 15], if ϑ is the handover signalling delay such that $P(t < \vartheta)$ is the probability that $t < \vartheta$, then the probability of false handover initiation, P_{False} is given by

$$P_{False} \begin{cases} 0, & \vartheta < \Phi_1 \\ P\left(t < \vartheta\right), & \Phi_1 < \vartheta < \Phi_2 \\ 1, & \vartheta > \Phi_2 \end{cases} \qquad (18)$$

For the range of values of ϑ in the interval $\vartheta_1 = \Phi_1 < \vartheta < \vartheta_2 = \Phi_2$ and using Equation (18), we obtain an expression for $P(t < \vartheta)$ as

$$P\left(t < \vartheta\right) = \int_t^\vartheta F_t\left(t\right) dt$$

$$= \int_{\vartheta_1}^{\vartheta_2} \frac{2r - \sqrt{3}r + 2\delta}{\theta_1 t\sqrt{\left[(2vt)^2 - \left(2r - \sqrt{3}r + 2\delta\right)^2\right]}} dt$$

$$= \frac{1}{\theta_1 \cos\vartheta\left[\frac{2r-\sqrt{3}r}{2v\vartheta}\right]} \qquad (19)$$

From Equations (18) and (19), we get the relationship between correct handover probability ($P_{correct}$) and vehicle's speed v, as follows.

$$P_{correct} \begin{cases} 0, & \vartheta > \Phi_2 \\ 1 - P\left(t < \vartheta\right), & \Phi_1 < \vartheta < \Phi_2 \\ 1, & \vartheta < \Phi_1 \end{cases} \qquad (20)$$

3.5 Cell Overlap Crossing Time and Data Throughput

For a given vehicle V, traversing an area covered by a wireless cell C at an average speed v, the cell crossing time of V through C, denoted by ΔT, is the overall time that V can spend under the coverage area of C [20]. From Figure 4, we can calculate the cell crossing time of the cell overlap region as follows. We assume the vehicle's trajectory follows a Manhattan mobility model and is constrained by straight lane roads. Let t_{in} and t_{out} denote the times at which vehicle enters the cell overlap region and the time at which it leaves the cell overlap region, respectively. The cell overlap crossing time is therefore, given by

$$\Delta T = t_{out} - t_{in}$$
$$= (2r - \sqrt{3}r + 2\delta)/(v cos\alpha) \quad (21)$$

where v is the vehicle's average speed. Therefore, the cell overlap region crossing time, ΔT is dependent on r, δ, α and v.

Given the above assumptions and definitions, we can model the data throughput, γ that the MN would experience by traversing through the cell overlap region during the period, ΔT as a function of the system bandwidth BW, which is assumed to be constant during the period ΔT. According to [20], the throughput experienced by the MN moving in the region comprising heterogeneous access radio cells during the period ΔT is a positive range function $\gamma : \Re \rightarrow \Re^+$ defined as

$$\gamma = \rho (B_{CN} - \eta)(\Delta T - T_L) + (1 - \rho) B_{SN} \Delta T \quad (22)$$

where ρ is an indicator function such that $\rho = 1$ when vertical handover is executed and zero otherwise, T_L is the handover latency, B_{CN} is the bandwidth of the candidate network, B_{SN} is the bandwidth of the serving network and $\eta \in \Re^+$ is the hysteresis factor introduced to avoid vertical handover occurrence when two competing networks have negligible bandwidth difference. In our model, we consider an intra-network handover scenario, hence a horizontal handover process. In this case, $\rho = 0$ and $B_{CN} = B_{SN} = BW$. The throughput due to the current BS is therefore given by

$$\gamma = \Delta T . BW$$
$$= \{ \left(2r - \sqrt{3}r + 2\delta \right) BW \} / (v cos\alpha) \quad (23)$$

where v is the average speed of the vehicle.

4 Simulations and Performance Analysis

4.1 Simulations Methodology

The implementation and simulations methodology are hereby presented in this section. Simulation input parameters used in the performance analysis are given in Table 1.

Simulations were conducted in the MATLAB numerical analysis tool environment. In the analysis, a scenario was considered where an IP capable MN moves from one subnet to another in a straight line at various average speeds in the range $(0 - 70 \; m/s)$. This is a typical range of speeds for most highway scenarios.

4.2 Performance Analysis

The performance evaluations were carried out in two phases. Firstly, we investigate the influence of RSS_{TH} position (hence, cell intrusion distance (δ)) and vehicle's average speed (v) on probability of correct handover initiation $(P_{correct})$. Secondly, we study the effects of vehicle speed (v) and direction (α) on data throughput (γ) during inter-subnet handovers.

4.2.1 Probability of Correct Handover Initiation ($P_{correct}$) Analysis

1. *Effects of RSS_{TH} Position on Probability of Correct Handover Initiation*

From Equation (10) representing the probability of correct handover initiation $(P_{correct})$, we can infer that if we unnecessarily increase the value of the cell intrusion distance (δ), $P_{correct}$ decreases. This results in wastage of the limited wireless system resources. Moreover, this increases the network load that arises due to handover initiation. Figure 5 shows the relationship between $P_{correct}$ and δ for different cell sizes, 'r'. The figure shows that, for a particular value of r, $P_{correct}$ decreases as δ increases. It can also be seen that the

Table 1   Simulation input parameters

Parameter	Symbol	Range of Values
Cell intrusion distance	δ	0–100 m
Vehicle's average speed	v	0–70 m/s
Vehicle's angular direction	α	$[-\pi, \pi]$ rad
Network bandwidth	BW	0.5–2.0 Mbps
Cell radius	r	1–10 Km
Handover signalling delay	ϑ	1–3 sec

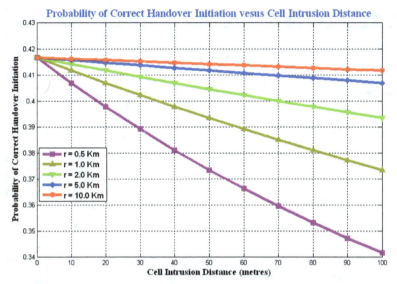

Figure 5 Relationship between ($P_{correct}$) and cell intrusion distance (δ)

problem of incorrect handover initiation becomes more and more pronounced when cell size decreases. Our main target in this respect is to increase $P_{correct}$. For this reason, we have to adjust the value of δ in such a way that $P_{correct}$ will be highest. In order to get a constant value of $P_{correct}$, we let $\delta = 0$, for which we get the value $P_{correct} = 0.417$.

2. Effects of Vehicle Speed on Probability of Correct Handover Initiation

From Equation (20), we can infer that when $\frac{2r-\sqrt{3}r+2\delta}{2v} < \vartheta < \frac{\sqrt{\frac{(2r-\sqrt{3}r+2\delta)^2}{4}+\frac{(\sqrt{3}r+2\delta)^2}{12}}}{v}$, for a fixed value of RSS_{TH} (and hence a fixed value of corresponding δ), the probability of correct handover initiation ($P_{correct}$) depends on the vehicle's speed. In fact, the probability of correct handover initiation decreases with increase in speed. That is, $P_{correct}$ is inversely related to vehicle speed. If speed is v, then we can write: $P_{correct} \propto \frac{1}{v}$. We testify this inverse proportionality with the help of a simulation. In our simulation, we considered a cell ($r = 3$ km) and the handover signalling delay ($\vartheta = 3sec$). Figure 6 shows the relationship between probability of correct handover initiation and vehicle's speed for an inter-subnet handover process. In the figure, numerical values of $P_{correct}$ for different values of δ (delta) which correspond to different values of RSS_{TH} positions are shown. Simulation results show that for a particular value of δ, as speed increases, $P_{correct}$

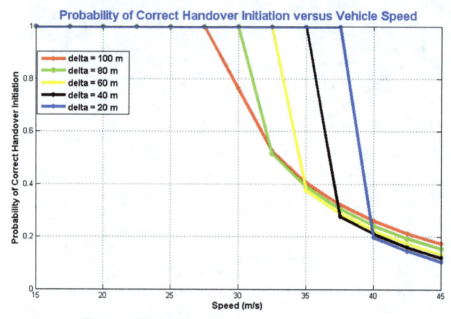

Figure 6   Relationship between $P_{correct}$ and vehicle speed

decreases since the vehicle requires less time to move out of the coverage range of the current BS into the coverage range of the target BS. Moreover, for a particular value of RSS_{TH} position, $P_{correct}$ becomes less when handover signalling delay is increased. This is usually the case for inter-subnet and inter-system handovers. On the other hand, intra-subnet handovers experience less signalling delays with typical values of less than 1 second in cellular networks such as UMTS systems [7, 15]. Therefore, for inter-subnet handovers to experience increased $P_{correct}$, ϑ must be reduced to values less than 1 second. However, this comes at a price of inaccurate IP configurations and registration processes.

4.2.2 Data Throughput (γ) Analysis

1. *Effects of Vehicle Speed on Data Throughput*

The relationship between data throughput, γ and vehicle speed, v is represented by Equation (23). From the equation, it can be deduced that throughput decreases as the average speed increases while the vehicle moves across the cell overlap region. Figure 7 shows the relationship between throughput due to current BS and vehicle's average speed for different cell sizes, 'r'

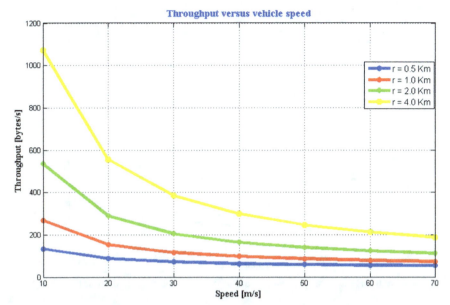

Figure 7 Relationship between data throughput and vehicle speed

at a constant network bandwidth of 2 Mbps. From the figure, it can be observed that for a single MN at a given speed, throughput is higher in a relatively bigger cell than a smaller cell. This is so because increasing cell size results in increased cell overlap region due to increased transmit power. Consequently, the MN takes longer time to handover when the serving radio cell is enlarged. To this end, we can state that increase in cell radius results in increased throughput. However, increased cell radius also results in increased interference from neighbour cells and this could compromise the network quality. Therefore, cell dimensioning should be carefully done so as to provide increased throughput while enjoying better network quality.

2. *Effects of Vehicle Direction on Data Throughput*

From Equation (23), it can be deduced that as the vehicle's angular direction, α is varied, the MN's data throughput, γ due to current BS varies. When α is increased with respect to line BI in Figure 4, γ increases since the vehicle spends more time within the coverage area of the current BS before it completely moves to the coverage area of the target BS. Numerical results depicting the relationship between data throughput and vehicle direction are

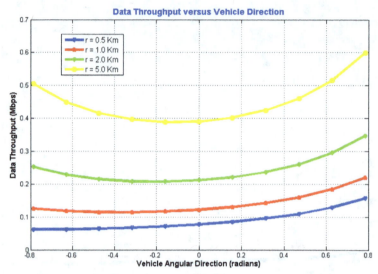

Figure 8   Relationship between data throughput and vehicle direction

shown in Figure 8. During simulations, the vehicle speed was fixed at a constant value of 45 m/s which is a typical highway speed. Numerical results show that throughput increases with cell size of the current BS for a particular value of α. However, typical cell sizes in heterogeneous access networks are in the range of 1–3 km [2, 7, 14]. Therefore, careful dimensioning of cell radii and positioning of RSS_{TH} (hence, δ value) must be taken into account during planning and designing of IP-based vehicular networks deploying throughput-sensitive applications.

5 Conclusions and Future Work

In this work, we introduced vehicular networks as an emerging wireless communication technology developed under the Intelligent Transportation Systems (ITS) to provide safety and non-safety related applications. We identified the negative effects of large signalling delays on throughput-sensitive and delay-intolerant ITS applications during inter-subnet handovers. This formed the motivation for this paper which led to the proposal of a mobility scheme aimed at reducing the signalling delay and increasing the data throughput by maximising the probability of correct handover initiation. Through numerical analysis, it is observed that for a fixed cell size value, the probability of correct handover initiation decreases as the value of RSS_{Th} position (hence, length δ) increases. Furthermore, when a fixed value of RSS_{Th} position is used, the probability of correct handover initiation decreases as vehicle speed increases. Based on this analysis, we suggest a method by which probability of correct handover initiation can be maximised and kept within constant limits. Furthermore, numerical analysis showed that data throughput decreases as vehicle speed increases. Moreover, data throughput increases with increase in vehicle's angular direction with respect to the shortest between the current BS and target BS during inter-subnet handovers. This gave us an insight on how cell size dimensioning can impact on data throughput and how a trade-off between data throughput and network quality has to be met during network planning and designing.

Future work should consider inter-Radio Access Technology (RAT) handovers comprising various heterogeneous access networks having different QoS demands and non-uniform coverage footprints. Furthermore, the proposed mobility model can be extended to standard vehicular network solutions such as the Wireless Access for the Vehicular Environment (WAVE) protocol stack for short-and medium-range V2V and V2I communications.

References

[1] S. Cespedes, X. Shen, C. Lazio, 'IP Mobility Management for Vehicular Communication Networks: Challenges and Solutions,' IEEE Communications Magazines, vol. 49, no. 5, pp. 187–194, May 2009.

[2] L. Banda, M. Mzyece, G. Noel, 'IP Mobility Support Solutions for Vehicular Networks,' IEEE Vehicular Technology Magazine (VTM), vol. 7, no. 4, pp. 77–87, Dec. 2012.

[3] Car-to-Car Communication Consortium, 'C2C-CC vehicular network architecture,' available at: http://www.car2car.org, 2011.

[4] V. Vassiliou, Z. Zinonos, 'An Analysis of Handover Latency Components in Mobile IPv6,' Journal of Internet Engineering, vol. 3, no. 1, pp. 230–240, 2009.

[5] L. Banda, M. Mzyece, G. Noel, 'An Enhanced FMIPv6 Scheme for Inter-subnet Handovers,' Southern Africa Telecommunication Networks and Applications (SATNAC), Spier Wine Estate, Western Cape, South Africa, 1–4 Sept. 2013.

[6] A. Chen, C. Wu, J. Ho, 'Secure transparent Mobile IP for intelligent transportation systems,' IEEE International Conference on Networking, Sensing and Controls, Taipei, Taiwan, 2004.

[7] D. Sarddar, et. al., 'Minimization of Handoff Failure Probability for Next-Generation Wireless Systems,' International Journal of Next-Generation Networks (IJNGN), vol. 2, no. 2, pp. 36–51, June 2010.

[8] M. M. Zonoozi, P. Dassanayake, 'User mobility modelling and characterisation of mobility patterns,' IEEE Journal on Selected Areas in Communication, vol. 5, no. 4, pp. 1239–1252, April 2011.

[9] S. Olariu, M. C. Weigle, 'Vehicular Networks: From Theory to Practice,' 1st ed., Taylor & Francis Group, Boca Raton, FL, 2009.

[10] F. Bai, N. Sadagopan, A. Helmy, 'The IMPORTANT framework for analysing the Impact of Mobility on Performance Of RouTing protocols for Ad-hoc NeTworks,' AdHoc Networks Journal-Elsevier Adhoc Networks, vol. 1, pp. 383–403, 2003.

[11] M. Fiore, J. Harri, 'The Networking Shape of Vehicular Mobility,' ACM MobiHoc, pp. 108–119, New York, NY, USA, Oct. 2007.

[12] M. Fiore, 'Mobility Models in Inter-Vehicle Communication Literature,' Tech. Rep., Nov. 2006 [Online]. Available: www.tlc-networks.polito.it/fiore/papers.

[13] L. Banda, M. Mzyece, G. Noel, 'An Analysis of Handover Probability and Data Throughput in Vehicular networks,' PACT 2013, Lusaka, Zambia, Jul. 2013.

[14] M. Al-Sanabani, et. al., 'Mobility Prediction Based Resource Reservation for Handoff in Wireless Cellular Networks,' International Arab Journal of Information Technology, vol. 5, no. 2, pp. 162–169, April 2008.

[15] D. Sarddar, et. al., 'Minimization of Handoff Latency by Angular Displacement Method using GPS Based Map,' International Journal of Computer Science Issues (IJCSI), vol. 7, no. 3, pp. 29–37, May 2010.

[16] H. Yokota, et. al., 'Link-layer assisted mobile IP fast handoff method over wireless LAN networks,' 8th Annual International Conference on Mobile Computing and Networking, pp. 131–139, 2002.

[17] C. Makaya, S. Pierre, 'Enhanced Fast Handoff Scheme for Heterogeneous Wireless Networks,' International Journal of Computer Communications, pp. 2016–2029, Jan. 2008.

[18] N. V. Hanh, S. Ro, J. Ryu, 'Simplified fast handover in mobile IPv6 networks,' Computer Communication, vol. 31, pp. 3594–3603, June 2008.

[19] L. Banda, M. Mzyece, G. Noel, 'Fast handovers without DAD using Sector-based Vehicular Mobile IPv6,' Southern Africa Telecommunication Networks and Applications (SATNAC), East London, South Africa, 4–7 Sept. 2011.

[20] F. Esposito, et. al., 'On Modelling Speed-Based Vertical Handovers in Vehicular Networks-Dad, slow down, I am watching a movie,' Annual IEEE Global Telecommunications Conference (GLOBE-COM '10), Miami, FL, USA, Dec. 2010.

Biographies

Laurence Banda received his BEng degree in electrical and electronic engineering from the University of Zambia (UNZA), Zambia in 2006, MSc degree in electronic engineering from ESIEE-Paris, France in 2011 and MTech degree in telecommunications engineering from Tshwane University of Technology (TUT), South Africa in 2012. He is currently working for Huawei Technologies, South Africa as a Wireless Trainer on LTE RNP and RNO products. His research interests include: vehicular networks, TCP/IP networks, 4G and beyond wireless broadband networks. (E-mail:laurencebandad@gmail.com).

Mjumo Mzyece is an associate professor with the French South African Institute of Technology (FSATI) and the Department of Electrical Engineering at Tshwane University of Technology (TUT), Pretoria, South Africa. He received a BEng (Honours) in electronic and electrical engineering from the University of Manchester, England, and an MSc (Distinction) in communications technology and policy and a PhD in electronic and electrical engineering, both from the University of Strathclyde, Scotland. (E-mail: mzyecem@tut.ac.za).

Impact of Topology on Energy Consumption in Wireless Sensor Networks

B. G. Awatef, N. Nejeh and K. Abdennaceur

National Engineering School of Sfax (ENIS), Sfax, Tunisia

Received 23 February 2014; Accepted 24 May 2014
Publication 4 August 2014

Abstract

Energy consumption is often regarded as a fundamental parameter in the context of availability in the wireless sensor networks (WSNs). This parameter poses energy problems particularly if the application must function for a long time. For the WSNs, it is impossible to reload or replace the batteries of nodes after their exhaustion. Several factors can be a source of energy over-consumption: mobility, retransmissions, the node position (relay or gateway), ... The network topology can also damage the nodes batteries. Generally, the transmission in 1-hop increases the energy consumption if the distance is rather high. The transmission in K-hops solves the problem but weakened the energy of the intermediate nodes. In this paper, we will focus on the impact of topology on energy consumption and determine optimal topology maximizing lifetime of WSNs.

Keywords: wireless sensor networks (WSNs), clustering, clusterhead, energy, topology.

1 Introduction

A WSN represents a set of small devices called sensors deployed in a geographical zone. The sensor is able to supervise a phenomenon and to send information to one or more point of collection inter-connected via a wireless connectivity ([1], [2]). Figure 1 represents the general diagram of

Journal of Machine to Machine Communications, Vol. 1, 145–160.
doi: 10.13052/jmmc2246-137X.124

WSNs: each sensor takes a physical greatness related to a noticed event, treat and transfer this data to a base station which transmits it towards the administrator.

Energy consumption (network lifetime) presents the most important metric in the performance evaluation of a WSNs ([3], [4]). Indeed, the lifetime is regarded as a fundamental parameter in the context of availability in the WSNs [5]. Energy consumption depends on several factors: state of the radio operator module (emission, reception or snooze for 802.15.4), retransmissions, idle listening, control packets, overhearing, collisions, ... [6]. The network topology represents also a major factor on energy consumption, indeed, a transmission in 1-hop requires a higher power if the distance between the nodes is high. A transmission in K-hops consumes less energy (for the initial node), but, can damage the batteries of the intermediate nodes.

Several researchers choose the transmission K-hops ([1, 7–11]) in order to minimize the transmission range. Others choose the transmission with 1-hop ([12–14]) without taking into account the distance between a source and a destination.

In this paper, we proposed an energy efficient clustering algorithm. A node with lower mobility, higher residual energy, higher degree (the number of nodes in its neighborhood of 1-hop) and closer to the base station is more likely elected as a clusterhead. Then, we want to determine optimal topology based on the clustering and minimizing energy consumption for the networks. Simulation results are described in the last section.

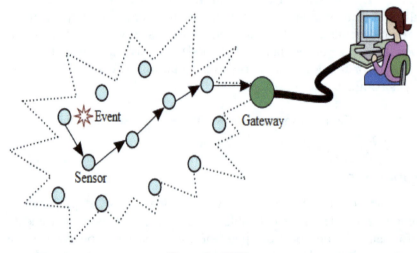

Figure 1   WSNs

2 Model and Notation

For better understanding the operation of a true WSN and thereafter facilitating its design, its deployment and its analysis, we will propose a model as simple as possible.

We model the WSN by a graph $G = (V, E)$ where V is the whole of sensors and $E = \{(u, v) \subset V2 \mid D (u, v) \leq R\}$ is the wireless connections between nodes. R is the communication range, and D (u, v) defines the Euclidean distance (the distance is calculated by the power of transmission) between the nodes u and v.

The properties are indicated in the Table 1.

3 MEDD-BS Clustering Algorithms

In this section, we propose a new clustering algorithm called MEDD-BS (Mobility Energy Degree Distance to Base Station) Clustering Algorithm for the wireless sensors networks of which the goal is the minimization of the energy consumption of WSNs.

3.1 Mobility Model

Mobility is the leading cause of topology changes in the sensors networks. It should be essential to integrate mobility metric for the clusterheads election and the clusters' formation.

There exist several applications where the nodes mobility is low like the mobility of a node deployed in the patient body to measure his physiological data. Other applications are characterized by a high nodes mobility, like the

Table 1 Parameters definition

N	Number of nodes.
ID (u)	Identifier of node u
D (u)	Connectivity degree of u.
M (u)	Mobility of u.
Ec/com (u)	Energy consumption by communication unit of u.
DisBS (u)	Distance between the node u and the base station.
Neigh (u)	Whole of nodes in the neighborhood of 1-hop of u.
D (u)	Degree of u.
Weight (u)	Weight of u.
State (u)	State of u. : "CH" (clusterhead) and "Nm" (nodesmember)
T	Period of standby mode (deactivation period).

mobility of a node deployed in the player body who runs in order to measure his speed.

So, we will define three mobility levels for sensors:

- Level 1: nodes speed is very weak in this case, speed varies between 0 and 5km/h.
- Level 2: nodes speed is average, in this case, speed varies between 5km/h and 20km/h.
- Level 3: nodes speed is high, in this case, speed varies between 20km/h and 44km/h.

We suppose that the sensors speed is constant. The sensor mobility is characterized by the mobility level and can have the following values:

- M (U) = 1, if the node speed U belongs to first level.
- M (U) = 2, if the node speed U belongs to second level.
- M (U) = 3, if the node speed U belongs to third level.

We also suppose that sensors nodes know in advance their mobility levels and that the nodes having meant and high mobility will not take part in the clusterheads election phase. The purpose of this assumption is to maintain the stability of the structure. We are interested in this paper by the nodes deployed in the human bodies, for that, the consumed power by the mobility of nodes is not considered into account.

3.2 Energy Consumption Model

The energy consumption rate in the sensors networks represents the most important metric in the performances "evaluation phase. This parameter depends on the used nodes" characteristics (standby mode, nature of data processing, transmitted power, ...), and nodes behavior during the communication (retransmission, congestion, diffusion of the messages, ...) [15].

The consumed power by sensor is that the consumed powers by these capture units, treatment units and communication units. So the energy consumption formula is defined as follows [16]:

$$Ec = Ec/capture + Ec/treatement + Ec/communication \quad (1)$$

Where:

- Ec/capture: is the energy consumed by a sensor during the capture unit activation. This energy depends primarily on the type of detected event (image, its, temperature...) and of the tasks to be realized by this unit (sampling, conversion...).

- Ec/treatment: is the energy consumed by the sensor during the activation of its treatment unit.
- Ec/communication: is the energy consumed by the sensor during the activation of its communication unit.

Generally, the consumed energy by sensors during communication is larger than those consumed by the treatment unit and the capture unit. Indeed, the transmission of a bit of information can consume as much as the execution of a few thousands instructions [17]. For that, we can neglect the energy of the capture unit, and the treatment unit compared to the energy consumed by the communication unit. In this case, the Equation (1) will be:

$$Ec = Ec/communication \qquad (2)$$

The communication energy breaks up into emission energy and reception energy:

$$Ec = E_{TX} + E_{RX} \qquad (3)$$

Referring to [18], the transmission energy and reception energy are defined as follows:

$$E_{TX} = E_{elec} \times K + \varepsilon_{amp} \times K \times d^{\lambda} \qquad (4)$$

$$E_{RX} = E_{elec} \times K \qquad (5)$$

Where:

- K: message length (bits).
- D: distance between transmitting node and receiving node (m).
- λ: of way loss exhibitor, $\lambda = 2$.
- Eelec: emission / reception energy, Eelec = 50 nJ/bit.
- ϵamp: transmission amplification coefficient, ϵamp = 100 pJ/bit/m^2

In [13], the authors compared the consumed power by a clusterhead by carrying out the aggregation of received messages with that consumed without aggregation. They showed that when the energy considered for aggregation is lower than a limit value (1µJ/bit/signal), then, the transmission with aggregation requires a weaker energy than that without aggregation.

We suppose that the aggregation energy cost respects the limit value introduced into [13]. The power consumed by a clusterhead during the transmission towards the base station will be:

$$E_{TX}(ch) = E_{DA} \times K + E_{TX}(K, d) \tag{6}$$

Where E_{DA}: power consumed during aggregation.

3.3 Clusterheads Election Procedure

Step 1: Each node sends a message "hello" for the discovery of 1-hop neighborhood.

Step 2: Nodes having a low level of mobility (M (u) = 1) calculate their weights, the weight is calculated as follows:

$$Weight(u) = Ec(u) + \frac{1}{D(u)} + \frac{DisBS(u)}{200} \tag{7}$$

We devide DisBS by 200 in order to take into account three parameters (three parameters will be between 0 and 1).

Step 3: The nodes diffuse their weights towards their neighbors (1-hop).

Step 4: The node which has the weakest weight is declared like clusterhead by putting its state = "CH" and sends a message "clusterhead_elected" (containing its identity) to its neighbors.

Step 5: The neighbors receiving this message, declare themselves like "Nm", send to the clusterhead a message "clusterhead_accepted", and record the identity of their clusterheads in their databases.

3.4 Clusterheads Change Procedure

This procedure is started if the waste energy of each CH will be lower than 40% of its initial energy.

Step 1: The clusterhead sends to its neighbors (1-hop) a message "clusterhead-changes", and is declared like "Nm".

Step 2: The nodes having a low mobility calculate and send their weights.

Step 3: The node having the weakest weight is declared like "CH", and diffuses a message "clusterhead_elected".

Step 4:   The neighbors send to the clusterhead a message "clusterhead_accepted", and record the identity of their clusterhead in their databases.

3.5 Organigram

The organigram of MEDD-BS algorithm is the following:

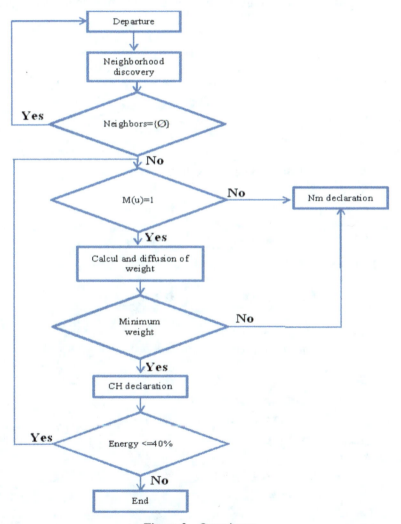

Figure 2   Organigram

4 Clusters Topology 1-HOP and K-HOPS

In the intra-cluster transmission, any member can be either 1-hop or k-hops of its cluster-head (Figure 3). In the cluster with 1-hop, the cluster-head is directly connected to any node member. In a cluster with K-hops, it is essential to define a procedure which calculates the way between node and its cluster-head.

The information routing towards the base station can be done into 1-hop (Figure 4) or in K-hops (Figure 5).

Thus, it is possible for a clustering strategy to define four topologies (Table 2).

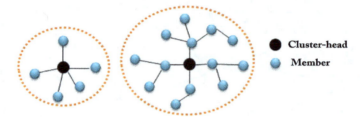

Figure 3   Connectivity intra-cluster

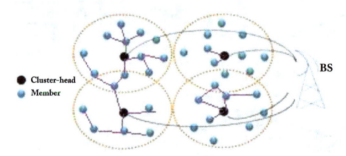

Figure 4   Connectivity inter-cluster 1-hop

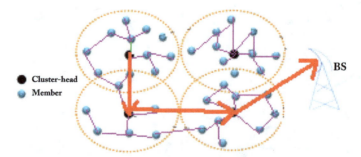

Figure 5   Connectivity inter-cluster k-hop

Table 2 Topologies

	1-hop Inter-cluster	K-hop Inter-cluster
1- hop Intra-cluster	Intra-cluster : 1-hop Inter-cluster : 1-hop	Intra-cluster : 1-hop Inter-cluster : k-hops
K-hops Intra-cluster	Intra-cluster : k-hops Inter-cluster : 1-hop	Intra-cluster : k-hops Inter-cluster : k-hops

5 Determination of the Optimal Topology

We wish thereafter to determine the topology optimizing energy consumption. For that, we will compare the power consumption in only one hop (E1) between a source S and a destination D and that consumed in two hops (E2).

Power consumption to directly transmit a message from S to D is equal to a transmission energy consumed by S and a reception energy consumed by D:

$$E1 = E_{elec} \times K + \varepsilon_{amp} \times K \times d3^{\lambda} + E_{elec} \times K \tag{8}$$

The Equation (8) becomes after simplification:

$$E1 = 2 \times E_{elec} \times K + \varepsilon_{amp} \times K \times d3^{\lambda} \tag{9}$$

Power consumption to transmit a message from S to D by G is equal to an energy of transmission consumed by S, an energy of reception consumed by G, an energy of transmission consumed by G and an energy of reception consumed by D:

$$E2 = E_{elec} \times K + \varepsilon_{amp} \times K \times d1^{\lambda} + E_{elc} \times K + \varepsilon_{amp} \times K \times d2^{\lambda} + E_{elec} \times K \tag{10}$$

The Equation (10) becomes after simplification:

$$E2 = 4 \times E_{elec} \times K + \varepsilon_{amp} \times K \times (d1^{\lambda} + d2^{\lambda}) \tag{11}$$

We want now to compare two energies E1 and E2, then, we will calculate E2-E1:

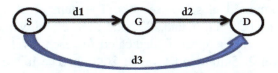

Figure 6 Transmission 1-hop and 2-hops

$$E2 - E1 = 2 \times E_{elec} \times K - \varepsilon_{amp} \times K \times 2 \times d1 \times d2 \qquad (12)$$

If E2-E1 is positive, then, the transmission in 1-hop is more profitable:

If:

$$E2 - E \geq 0 \qquad (13)$$

Then:

$$2 \times E_{elec} \times K - \varepsilon_{amp} \times K \times 2 \times d1 \times d2 \geq 0 \qquad (14)$$

After simplification, we will have:

$$d1 \times d2 \leq \frac{E_{elc}}{\varepsilon_{amp}} \qquad (15)$$

Then:

$$d1 \times d2 \leq 500 \ m^2 \qquad (16)$$

So, the transmission in 1-hop is profitable when d1*d2 $\leq$ 500 and the transmission in 2-hops is profitable when d1*d2 $\geq$ 500 m^2.

Within a cluster (transmission intra-cluster), the distances between the nodes are generally weak (about 10m and 20m), then, the multiplication between two distances is lower than 500, which favour the transmission in 1-hop.

For the inter-cluster transmission, the distances between nodes are generally larger (about 70 m and 100 m), then, the multiplication between two distances is higher than 500, which favour the transmission into 2-hop.

So, we can conclude that the optimal topology is topology 1_k (1-hop intra-cluster and K-hops inter-cluster).

6 Simulation Results

The simulation results were implemented using Matlab 7.0.1 tool. The network of first level is composed of set of sensors. The number of nodes in the sensor network varies between 10 and 200 nodes. The mobility of each sensor is supposed constant, and a speed is dedicated for each level of mobility: level 1: 1km/h, level 2: 5km/h and level 3: 20km/h. The initial energy for each sensor is equal to 0.5 J.

The simulation of the proposed algorithm was carried out during 10 intervals T (standby mode) in a space of 150 m × 150 m and the range of the nodes (Tx-Arranges) is 40 m. The size of a measured data package for sensors and envoy towards their clusterheads is 4000 bits. Figure 7 shows the communication structure of network with 30 nodes. In this figure, red o represent the cluster-head, yellow triangle represent the ordinary sensor node, blue lines represent the communication between cluster-head and ordinary sensor nodes, and black * represent the base station.

In this section, we will represent the results of our algorithm by varying the transmission topology. Figure 8 improves the results shows in the preceding

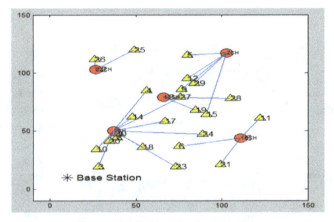

Figure 7　Communication structure

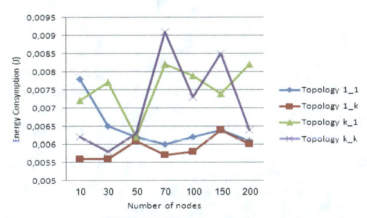

Figure 8　Energy consumption

part. Topology 1-k is topology which minimizes energy consumption for WSNs.

Figure 9 represents the medium number of constructed CHs VS number of nodes. We can notice that four topology give comparable values of CHs. The transmissions 1-hop between members and its CHs increases the number of the built clusters. The transmissions k-hops increases the number of nodes in the cluster and, by consequence, the number of constructed cluster decrease.

Figure 10 represents the average number of transmitted packets to CHs. "Topology 1_k" and "Topology k_k" gives the highest values. For "Topology

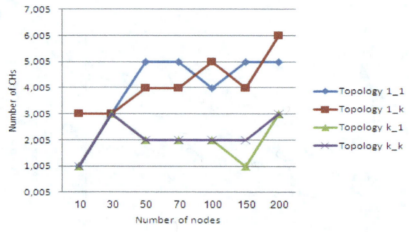

Figure 9   Number of CHs

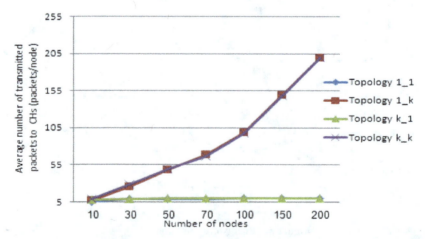

Figure 10   Transmitted packets to CHs

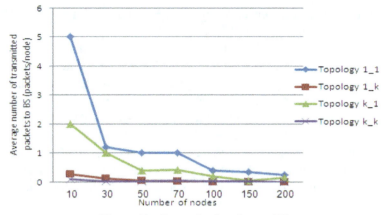

Figure 11   Transmitted packets to BS

1_k", a large number of CHs is remarked what makes the packages value higher. For "Topology k_k", a way multi-hops is carried out at the time of transmissions towards the BS, thus more number of packages are sent between CHs.

Figure 11 represents average number of transmitted packets to the BS. The lowest values are those of "Topology 1_k" and "Topology k_k", only one CH carries out the procedure of transmission towards the base station.

7 Conclusion

In this paper, we focussed on the impact of the topology on the energy consumption in the WSNs. Theoretical and simulation results have shown that the topology 1_k (1-hop intra-cluster and k-hop inter-cluster) optimize energy consumption and increases by consequence the network lifetime. The next step should be the construction of a topology 1_K which minimize the energy consumption of each node in the WSN.

References

[1] C. Tidjane Kone, "Conception de l"architecture d"un réseau de capteurs sans fil de grande dimension", PHD, University of Henri Poincaré Nancy I, October 2011.

[2] G. Chalhoub, "Les réseaux de capteurs sans fil", Ph.D, University of Clermont, Auvergne, 2010.

[3] L. Samper, "Modélisations et Analyses de Réseaux de Capteurs", PHD, VERIMAG laboratory, April 2008.

[4] R. Kacimi, "Techniques de conservation d"énergie pour les réseaux de capteurs sans fil", PHD, TOULOUSE UNIVERSITY, September 2009.

[5] M. Khan and J. Misic. "On the lifetime of wireless sensor networks", ACM Transactions on Sensor Networks (TOSN), Vol. 5, No. 5, 2009.

[6] R. Kuntz, "Medium access control facing the dynamics of wireless sensor networks", PHD, University of Strasbourg, 2010.

[7] D. Kumar, D. Trilok, C. Aseri and R. B. Patel, "EEHC: Energy efficient heterogeneous clustered scheme for wireless sensor networks", *Computer Communication.*, Volume 32, Issue 4, pp 662–667, 4 March 2009.

[8] W. B. Heinzelman, "Application-Specific Protocol Architectures for Wireless Networks", PHD, B.S.Cornell University, June 2000.

[9] Y. Ossama and S. Fahmy, "HEED: A hybrid energy-efficient distributed clustering approach for ad hoc sensor networks", IEEE Transactions on Mobile Computing, Volume 3, Issue 4, pp 366 – 379, 2004.

[10] Duan, Changmin et Hong Fan, "A distributed energy balance clustering protocol for heterogeneous wireless sensor neworks", In: Proc. Int. Conf. Wireless Communications, Networking and Mobile Computing WiCom, pp. 2469–2473, 2007.

[11] Zhenhua Yu, YuLiu and Yuanli Cai, "Design of an Energy-Efficient Distributed Multi-level Clustering Algorithm for Wireless Sensor Networks", Proc IEEE 4th International Conference: Wireless Communications, Networking and Mobile Computing (WiCOM"08),2008.

[12] T. Neeta, G. Elangovan, S. Iyengar and N. Balakrishnan, "A message-efficient, distributed clustering algorithm for wireless sensor and actor networks", IEEE Int Multisensor Fusion and Integration for Intelligent Systems Conf, 2006.

[13] Wendi Rabiner Heinzelman, Anantha Chandrakasan, and Hari Balakrishnan, "Energy-Efficient Communication Protocol for Wireless Microsensor Networks", In : Proc IEEE 33rd Hawaii International Conference on System Sciences, 2000.

[14] CP. Low, C. Ping, C. Fang, J. Mee and Y. Hock Ang, "Load-balanced clustering algorithms for wireless sensor networks", Proc. IEEE Int Conf Communications ICC, 2007.

[15] V. Raghunathan, C. Schurgers, S. Park and M.B. Srivastava, "Energy-aware wireless micro-sensor networks", IEEE Signal Processing Magazine, Vol. 19, No. 2, pp 40–50, 2002.

[16] Mitton, Nathalie, Busson, Anthony et Fleury, Eric. "Self-organization in large scale adhoc networks", In : Mediterranean ad hoc Networking Workshop (Med-Hoc- Net"04). Bodrum, Turquie, 2004.
[17] G. J. Pottie and W. J. Kaiser, "Wireless integrated network sensors", Commun. ACM, Volume 43 Issue 5, pp 51–58, May 2000.
[18] W. N. Richard and A. Boukerche, "Mobile data collector strategy for delay - sensitive applications over wireless sensor networks," Computer Communications, Volume 31, Issue 5, pp 1028–1039, 25 March 2008.

Biographies

Awatef BENFRADJ was born in Gabes, Tunisia in 1983. She received the engineering degree in Telecommunications and Networks from the University of Gabes, in 2007. In 2009, she obtained the master degree in Networks from Higher School of Communication of Tunisia. Since 2011, she has been a researcher within the laboratory of electronics and information technology, Sfax University.

Nejah NASRI was born in Menzel Bouzaienne Village, in 1981. He received the B.S. and M.S. degrees in electronic engineering from the University of Sfax, in 2006 and the Ph.D. degree in electronic and telecommunication engineering from Toulouse University, France, in 2010. From 2006 to 2009, he was a Research Assistant with Higher Institute of Computer and Communication

Techniques (ISITCom), Hammam Sousse, Tunisia. Since 2010, he has been an Assistant Professor with the Informatics Engineering Department, Gafsa University. He is the author of more than 70 articles. His research interests include engineering of wireless sensors networks and wireless underwater communication.

Abdennaceur KACHOURI was born in Sfax, Tunisia, in 1954. He received the engineering diploma from National school of Engineering of Sfax in 1981, a Master degree in Measurement and Instrumentation from National school of Bordeaux (ENSERB) of France in 1981, a Doctorate in Measurement and Instrumentation from ENSERB, in 1983. He "works" on several cooperation with communication research groups in Tunisia and France. Currently, he is Permanent Professor at ENIS School of Engineering and member in the "LETI" Laboratory ENIS Sfax.

Designing Smart Homes for Dependent Persons Assistance

Rim Jouini, Karima Maalaoui and Leila Azouz Saidane

Ecole Nationale des Sciences de l'Informatique, University of Manouba Campus of Manouba, Manouba, Tunisia

Received 23 February 2014; Accepted 24 May 2014
Publication 4 August 2014

Abstract

Evolution of technology has permitted to help dependent persons by implementing new systems to assist them at their homes. The majority of these systems are based on Wireless Sensor Networks (WSN) using 802.15.4 standard at the MAC level. In this paper, we propose a solution to permit a continuous assistance for dependent persons, called Domotic Assistance for Dependent Persons (DADP). We also propose a new MAC protocol to enhance QoS requirements, especially for alarm management in critical situations. Simulation results have shown that our proposed solution performs better than 802.15.4 in terms of reliability, energy consumption and loss ratio.

Keywords: WSN, IEEE 802.15.4.

1 Introduction

Recent years has known a continuous increase in the number of dependent persons. A more simple and transparent hardware, attentive and adaptable to the user lifestyle, is still an open issue [11]. To deal with this issue, we propose an integrated and low-cost wireless architecture for noninvasive monitoring. WSN is an example of a technological solution adopted by researchers to go toward the idea of ubiquitous computing and smart

Journal of Machine to Machine Communications, Vol. 1, 161–176.
doi: 10.13052/jmmc2246-137X.125

environment [11]. These networks are formed by a set of miniature nodes that collect information and send it to a configured sink node that forwards it to a monitoring system. A sensor node is composed by a sensing unit for acquiring data, a micro-controller for data computation, a wireless communication unit and a limited power battery. WSN will permit a continuous assistance for dependent people in their homes as opening doors and turning off or on the light automatically. This needs real time traffic to respond to dependent persons needs in a real time manner. In the MAC level, IEEE 802.15.4 standard is considered as the most widely adopted solution for WSNs [1, 2]. However, this standard cannot support application with QoS requirements and is not suitable for real time traffic management [3]. To solve this problem, we have defined a new MAC protocol under LEACH. This new protocol will enhance QoS requirements especially for alarm management in critical situations by providing them with high priority while ensuring low energy consumption and high reliability13 The rest of the paper is organized as follows. The first section gives an overview of the IEEE 802.15.4 protocol. The second section presents our solution, called DADP, for assisting dependent persons in their homes. Performance evaluation of DADP is presented in the third section. Finally, we conclude the paper and outline our future work.

2 IEEE 802.15.4 Protocol

The following section gives a brief overview of IEEE 802.15.4 MAC protocol. This protocol is defined to provide applications with low throughput and latency requirements. Besides, low complexity, low cost, low power consumption, low data rate transmissions are the key characteristics of its adoption for WSNs [2]. IEEE 802.15.4 supports three topologies: star, tree or mesh topology. Two types of devices can coexist: FFD (Full-Function Devices) which implement all the functions of the standard and RFD (Reduced-Function Devices).

IEEE 802.15.4 MAC has two operational modes: a beacon enabled mode and a non-beacon enabled mode. In the first mode, the network is controlled by a coordinator, which regularly transmits beacons for device synchronization and association control. The channel is bounded by a superframe structure as illustrated in Figure 2. The superframe consists of both active and inactive periods. The active period contains three components: a beacon, a Contention Access Period (CAP) and a Contention Free Period (CFP). The coordinator interacts with nodes during the active period and sleeps during inactive period. There are a maximum seven Guaranteed Time Slots

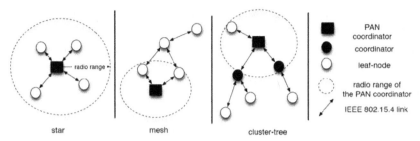

Figure 1   The different topologies proposed in IEEE 802.15.4

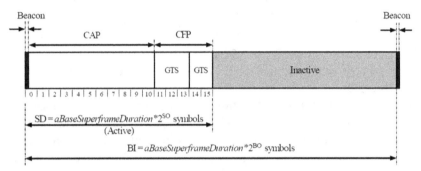

Figure 2   IEEE Superframe [4]

(GTS) in the CFP period to support time critical traffic. In the beacon-enabled mode, a slotted CSMA/CA protocol is used in the CAP period while in the non-beacon enabled mode, unslotted CSMA/CA protocol is used [4].

3 Related Works

An aging population, combined with sedentary lifestyle and poor diet, is resulting in an increasing number of dependent people that require help. Having a person as a helper at home is very expensive. Wireless sensor network technologies have the potential to offer large scale and cost effective solutions to this problem. They permit in-home monitoring of elderly, patients or disabled persons. Such technologies have enabled remote monitoring of dependent persons in their own homes or environment and improved care provider ability to deliver help.

WSN have to send periodic message to a monitoring center by conducting their messages to a specified access point. Abnormal critical situation cases should have a special treatment to insure timely delivery for efficient intervention.

Following the messages they have to undertake, two types of traffic can be found in a WSN:

- Periodic traffic indicates the best function of node
- Alarms generated in critical situations.

These messages need real time and high priority treatment.

In the literature, several solutions have been proposed for dependent persons assistance [10] as GRADIEN, PROSAFE, EVIENT, ERGDOM and AILISA. Next, we will focus only on the two first solutions as they are the most important of all solutions. GRDEIEN was developed by INSERM U558 and ProSafe system was developed by LAAS in Toulouse, to monitor the nocturnal wanderings in a hospital room, called Hospital Smart Rooms. GARDIEN aimed to trigger an alarm if the patient quits the room and help diagnose symptoms of a disease through the study of motor behavior and appreciate the effects of treatment on motor behavior of the patient. PROSAFE permits also the same type of remote monitoring of frail patients, but it improves the previous system using wireless sensors, but it proposes a system of automatic alarm triggering based on various criteria.

These different projects were developed for dependent person aim, especially for monitoring health. Our idea is to look for this dependent person from another sight that consists to assist this person in their own homes or environment to facilitate their activities and improved care providers ability to deliver help in critical situations.

4 Idea of Smart Home

Smart home integrates information and communication-technology, whereby the different components communicate via a local network. This network communicates with the external world by telephone or through the Internet, which means that the smart home can be programmed from inside or outside the house.

Smart home technology can contribute to increased independence and safety for disabled person. Safety is the biggest advantage of smart home technology. Smart home is advertised to call customers as a means of living more comfortably in their home. To the disabled person, this advantage means an increased ability to cope with the activities of daily living, resulting in increased independence.

5 DADP Solution

We propose a novel solution called Domotic Assistance for Dependent Person (DADP) and a new medium access control protocol to manage alarms and periodic messages. DADP will permit assistance for dependent persons in their habitat as shown in Figure 3.

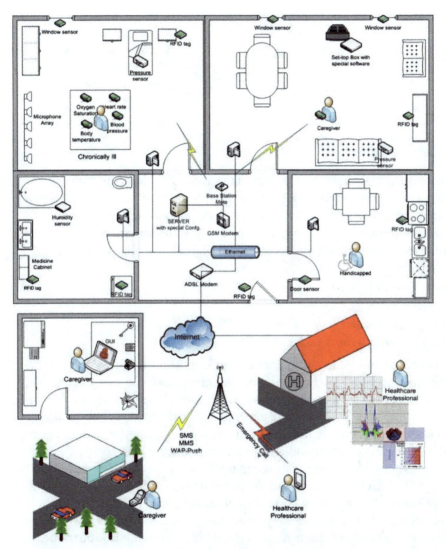

Figure 3 WSN overview in a sample smart home for person assistance [5]

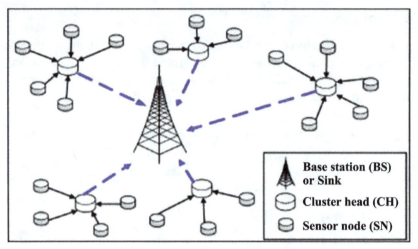

Figure 4   Cluster-based WSN [6]

To ensure the safety of a dependent person, we place, in each room, a gas detection sensor node and a smoke detection sensor node. These sensors generate alarms that are transmitted to the center monitoring or referred person. The actuators nodes are also placed in every room to manage brightness, window pane and temperature as shown in Figure 2. Actuators of our DADP solution are:

- Gas sensor nodes.
- Smoke sensor nodes.
- Light sensor nodes.
- Temperature sensor nodes.

These elements generate a periodic traffic and alarms. The periodic traffic permits to verify the good function of the sensor nodes in every room. This traffic is received by a room coordinator sensor (Room head) and will be relayed to the Sink. The Alarms are generated upon detection of smoke or gases and it should be sent immediately in a short time to have an early intervention.

All sensor and actuator nodes form a communication network in the smart home. This network can be organized in two topologies: a Cluster-based WSN topology (CWSN) or a Flat-based WSN topology (FWSN).

In FWSN, the multi-hop communication generates an important flow of information and is high level energy consuming. Thus, the control traffic generates a high overhead in the detriment of the effective communication traffic [6]. To solve this problem, one solution could be dividing the network into clusters.

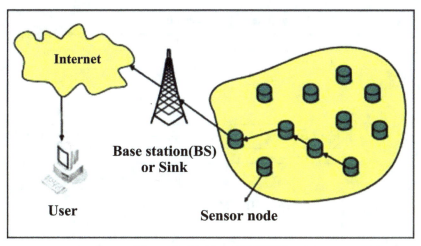

Figure 5   Flat-based WSN [6]

For CWSN, geographically nearby sensor nodes form a cluster identified by a particular node among them called cluster-head (CH). The CH manages its cluster and aggregates the traffic inside the cluster before sending it to the sink. Thus, both overhead and energy consumption are reduced. Many protocols as CWSN have been proposed as LEACH APTEEN and Pegasus [6].

To organize the wireless sensor network in the home of a dependent person, we adopt the cluster star architecture by using the protocol LEACH. We propose a new access method to overcome the limitations of the access standard IEEE 802.15.4 MAC layer. This method will be used to enhance the management of alarms and periodic messages. It relies on the static nature of the home network to implement a fixed scheduling strategy. It will give to sensors reporting urgent information a higher priority by delaying transmission of packets that are reporting periodic low priority messages. The proposed MAC protocol be used with LEACH protocol.

In the same context, Hanen and al. [7] have proposed a novel medium access control protocol for communication in a WBAN for remote monitoring of physiological signals of patients by implementing a fixed scheduling strategy in which they keep the same concept of beaconing as used in IEEE 802.15.4 but they change the decomposition of the superframe that they called E-Health Access Channel Mechanism (EHACM).

They divide the E-HACM superframe into equal Allocated Time Slots (ATSs) (see Figure 6). Each ATS is allocated to a node according to a static

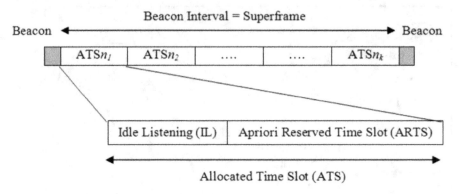

Figure 6   E-HACM Superframe [7]

predefined scheduling. Each node can be in sleep mode and wakes up only during its allocated ATS or if it has an alarm. An ATS contains two sub periods: Idle Listening (IL) and Allocated Reserved Time Slot (ARTS). IL is dedicated for sending alarms if any node of the network needs to, while the ARTS period is reserved to the owner of the current ATS to send its periodic message if no alarm was generated during the IL period of the current ATS. The owner of the current ATS has to listen during IL and cancel its periodic sending if another node has generated an alarm during the IL period. Once a node has completed its sending, it returns into sleep mode until the next superframe.

In our solution DADP, the MAC protocol (Figure 7) is inspired from E-HACM. In addition, the IEEE 802.15.4 superframe is divided into 16 slots, each one is assigned to a node. Our contribution is limited only to MAC layer using the IEEE 802.15.4 default physical layer. That's why we limit ourselves to 16 slots in our solution. The general idea is to divide the superframe into equal Cluster Allocated Time Slots (ATSs-Cluster) (Figure 3). Each ATS-Cluster is divided into equal ATS-Node according to the E-HACM method. Each node can be in sleep mode and wakes up only during its allocated ATSNode or if it has an alarm. An ATS-Node contains two sub periods: Idle Listening (IL) and Allocated Reserved Time Slot Node (ARTS-Node). IL is dedicated for sending alarms if any node of the network needs to, while the ARTS-Node period is reserved to the owner of the current ATS-Node to send its periodic message if no alarm was generated during the IL period of the current ATS-Node. The owner of the current ATS-Node has to listen during IL and cancel its periodic sending if another node has generated an alarm during

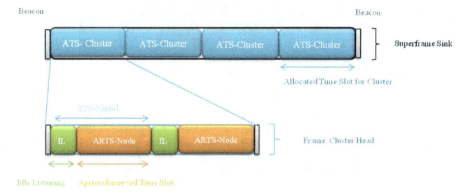

Figure 7   DADP Superframe

the IL period. Once a node has completed its sending, it returns into sleep mode until the next superframe.

6 Performance Evaluation

We implemented DADP in Castalia [8] under OMNET++ simulator [9]. We have evaluated the performances of DADP according to the following criteria:

- The MAC level packet breakdown
- The number of packets received per sink
- The consumed energy
- The loss ratio

We run the same simulation scenarios for IEEE 802.15.4 with its two modes (Beacon and Beaconless modes) in order to compare them with the performances of our proposed solution while varying the rate. Simulation results are shown in the following subsections.

A. MAC level packet breakdown

Figure 8 shows the distribution of packets at the MAC level "break Packet" while varying rate. We notice no packets failure because access is based on reserved slots.

B. Number of packet received

For IEEE 802.15.4 beacon mode, the number of packet received by sink is limited because we have a maximum of 7 GTSs in an IEEE 802.15.4 superframe. Thus, some nodes will cancel their messages since they will not have allocated GTS.

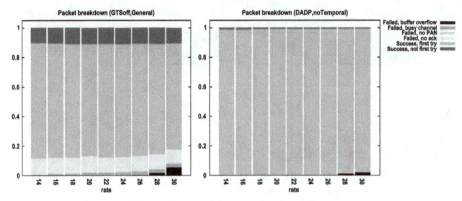

Figure 8 MAC level packet breakdown

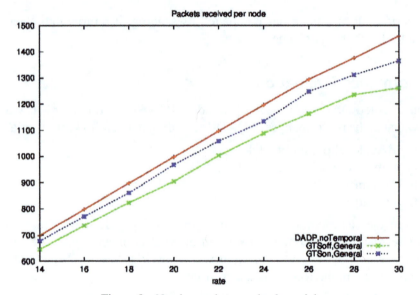

Figure 9 Number packets received per sink

C. Energy consumption

Figure 10 shows energy consumption while varying rate. We notice that, although the values of the two protocols are almost the same, DADP is more energy efficient compared to IEEE 802.15.4 due to the static scheduling defined in DADP. Nodes spend more time in sleep state, thus conserving their energy. The measures obtained by GTSoff can be explained due to use of the method CSMA/CA.

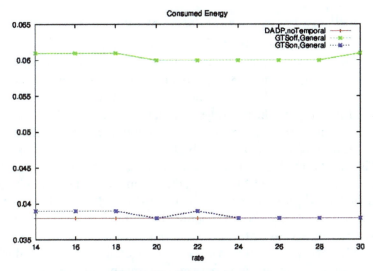

Figure 10 Consumed energy

D. Loss ratio

Figure 11 shows the loss ratio generated by DADP. We notice that the majority of packets are successfully sent for the first try. Therefore, DADP is more suitable for a real time transfer and is more reliable for alarms management.

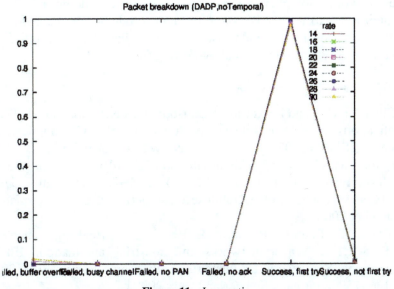

Figure 11 Loss ratio

7 Conclusion

In this paper, we proposed a WSN based solution for dependent persons assistance at their smart homes. This solution, called DADP, permits to a distant manager to take care of a dependent person at his smart home by communicating with the WSN. We have also introduced a novel MAC protocol to manage alarms and periodic messages generated by sensors. This protocol relies on the static nature of the home network to implement a fixed scheduling strategy. The new MAC protocol insures high reliability for critical situations.

DADP is based on a fixed scheduling mechanism where each sensor has a pre-assigned time slot, called ATS-Node, during an ATS-Cluster. Each node should be awake only during its preassigned ATS-Node. Nonetheless, in case of critical situations, a sensor should generate immediately an alarm. It can use any ATS-Node of ATS-Cluster for its alarm message transmission. To avoid collisions, each ATS-Node begins with an idle listening sub period, during which any sensor handling an alarm message can transmit immediately, while the assigned node has to listen before deciding to cancel or to send its periodic message.

Our solution has been validated through simulations using Castalia and OMNET++ simulator. We have proved that the proposed solution enhances reliability transmission, energy consumption and loss ratio compared to IEEE 802.15.4.

In future works, we plan to extend our solution by adding a WBAN as dynamic cluster to the DADP static clusters. Modeling the behavior of the dependent person in its habitat using Markov chain for more efficient consumption energy may be a second interesting future work.

References

[1] B. Zhen, Li. H-B and R. Korno. IEEE Body Area Networks for Medical Applications. The International symposium on Wireless Communication Systems (ISWCS 2007), pp. 327–331, Norway, October 2007.

[2] P. Kumar, M. Gunes, A.B. Almamou and J. Schiller. Real-time, Band width and Energy Efficient IEEE 802.15.4 for Medical Applications. 7th GI/ITG kuVS Fachgesprach Drahtlose Sensornetze, FU Berlin, Germany, September 2008.

[3] F. Chen, T. Talanis, R. German, F. Dressler. Real-time enabled IEEE 802.15.4 sensor networks in industrial automation. IEEE International Symposium on Industrial Embedded Systems (SIES'09), pp. 136–139, Lausanne, Switzerland, July 2009.

[4] S. Ullah, H. Higgins, B. Braem, B. Latr, C. Blondia, I. Moerman, S.Saleem, Z. Rahman, and K.S. Kwak, "A Comprehensive Survey of Wireless Body Area Networks - On PHY, MAC, and Network Layers Solutions", Springer: Journal of Medical Systems, pp.1065–1094, June 2012.

[5] H. Alemadr and C. Erosy, "Wireless Sensor Networks for Healthcare: A Survey" ScienceDirect: Journal of Computer Networks, Volume 54, Issue 15, pp.2688–2710, October 2010.

[6] Shun-Sheng Wang, Kuo-Qin Yan, Shu-Ching Wang, Chia-Wei. An Integrated Intrusion Detection System for Cluster-based Wireless Sensor Networks. Journal Expert Systems with Applications: An International Journal, Volume 38, Issue 12, pp. 15234–15243, November, 2011.

[7] Hanen Idoudi, Fatma Somaa, Leila Azouz Saidane. "Alarms Management in Wireless Body Area Networks". International Conference on Computer Systems and Applications (AICCSA), 2013 ACS, pp. 1–4, France, May 2013.

[8] Castalia simulator for Wireless Sensor Networks (WSN), Body Area Networks(BAN) and generally networks of low-power embedded devices. http://castalia.research.nicta.com.au/index.php/en/. Last access on June 2013.

[9] OMNET++ Network Simulator. http://www.omnetpp.org/. Last access on June 2013.

[10] M. Chan, E. Campo, D. Estve, PROSAFE, a multisensory remote monitoring system for the elderly or the handicapped in Independent Living for Persons with Disabilities and Elderly People, IOS Press,1st International Conference On Smart homes and health Telematics (ICOST'2003), Sept. 24 26, 2003, pp. 89–95, Paris France.

[11] M. R. Alam, M. B. I. Reaz, and M. Ali, "A review of smart homes Past, present, an future," IEEE Trans. Syst. Man Cybern. C, Appl. Rev., vol. 42, no. 6, pp. 1190–1203, Nov. 2012.

Biographies

Rim JOUINI member and researcher of CRISTAL - ENSI lab Tunisia, received the diploma of national computer engineering from ENSI, University of Manouba Tunisia in 2013, and MS degree from ESTI, University of Carthage Tunisia in 2010 and bachelor degree in Computing from *ISET of* RADES 2008.

Karima MAALAOUI is an assistant professor at Carthage University's Faculty of Sciences. She received the Engineer degree in 2003 and The Ph. D. in 2009, all from the National School of computer sciences-university of Manouba. Her areas of interest and research include QoS and security in wireless and mobile networks, M2M communications, Internet of things and Wireless sensors networks.

Leila Azouz Saïdane is Professor at the National School of Computer Science (ENSI), at The University of Manouba, in Tunisia and the Chairperson of the PhD Commission at ENSI. She was the Director of this school and the supervisor of the Master's Degree program in Networks and Multimedia Systems. She is the co-director of RAMSIS group of CRISTAL Research Laboratory (Center of Research in Network and System Architecture, Multimedia and Image Processing) at ENSI. She collaborated on several international projects. She is author and co-author of several papers in refereed journals, magazines and international conferences.

An Adaptive TDMA Slot Assignment Strategy in Vehicular Ad Hoc Networks

M. Hadded[1], R. Zagrouba[2], A. Laouiti[3], P. Muhlethaler[4], and L. A. Saïdane[5]

[1]*RAMSIS Team, CRISTAL Laboratory, 2010 Campus University, Manouba, Tunisia*
[2]*Higher Institute of Computer Science, Ariana, Tunisia*
[3]*TELECOM SudParis, CNRS Samovar, UMR 5157, France*
[4]*INRIA, BP 105. 78153 Le Chesnay Cedex, Paris-Rocquencourt, France*
[5]*National School of Computer Science, 2010 Campus University, Manouba, Tunisia*

Received 23 February 2014; Accepted 25 May 2014
Publication 4 August 2014

Abstract

Improving road safety is among the main objectives of Vehicular Ad-hoc NETworks (VANETs) design. This objective would be achieved essentially by the use of efficient safety applications which should be able to wirelessly broadcast warning messages between neighbouring vehicles in order to inform drivers about a dangerous situation in a timely manner. To insure their efficiency, safety applications require reliable periodic data dissemination with low latency. Medium Access Control (MAC) protocols play a primary role to provide efficient delivery and avoid as much as possible data packet loss. In fact, in distributed MAC approaches, packet loss is a consequence of collisions resulted from well known situations of the exposed and hidden node situations. This paper introduces an Adaptive TDMA Slot Assignment Strategy (ASAS) for VANET based on clustering of vehicles. The main aim of this work is to provide a MAC layer protocol that can reduce inter-cluster

Journal of Machine to Machine Communications, Vol. 1 , 177–196.
doi: 10.13052/jmmc2246-137X.126

interference under different traffic loading conditions without having to use expensive spectrum and complex mechanisms such as CDMA or OFDMA. An analysis and simulation results are presented to evaluate the performance of ASAS. Moreover, we compare its performance with two TDMA MAC protocols DMMAC and VeMAC.

Keywords: VANET, QoS, MAC protocols, CDMA, TDMA, FDMA.

1 Introduction

The Vehicular Ad-hoc NETwork (VANET) is a sub class of Mobile Ad-hoc NETwork (MANET), which has some special characteristics such as the high dynamicity of the nodes, the lack of infrastructure and diverse quality of service (QoS) requirements. In VANETs, communications can either be between nearby vehicles V2V (Vehicle To Vehicle) or between vehicles and road side units V2I [9] (Vehicle To Infrastructure). Due to the importance of V2V communications, several research projects are underway to standardize V2V communication in Europe and around the world such as the Car2Car consortium [5] which seeks to improve road safety. In the US, the FCC (Federal Communication Commission) [2] established the DSRC service (Dedicated Short Range Communications) in 2003. The DSRC [3] radio technology is defined in the frequency band of 5.9 GHz with a total bandwidth of 75 MHz. This band is divided into 7 channels of 10 MHz for each one. These channels comprise one Control CHannel (CCH) and six Service CHannels (SCHs), each one offering a throughput from 6 to 27 Mbps. The CCH is reserved for the network management messages, but is also used to transmit messages of high priority. The six SCHs are dedicated to data transmission.

However, the V2V communication is based on the exchange of beacon messages (current status, aggregate data, and emergency messages). If several vehicles broadcast these messages at the same time, then a collision will occur. Thus, it is crucial to avoid collision on the Control CHannel CCH in order to ensure a fast and reliable safety messages exchange. To provide a QoS and reduce the collision on the CCH, we introduce an adaptive TDMA slot allocation strategy that takes into account the specificity of VANETs networks. The strategy proposed operates at the stable clusters heads which are responsible for assigning disjoint sets of time slots to the members of their clusters according to their directions and positions. Thus, by using a centralized means of slots reservation, we ensure an efficient utilization of the time slots and thereby decrease the rate of merging collision [13] and hidden node collisions caused by vehicles moving in opposite directions.

The remainder of this paper is organized as follows. Section 2 reviews related works on MAC protocols in VANET. Section 3 sets out the challenges of TDMA based MAC solution deployment. We give a detailed description of ASAS in Section 4. In Section 5, we present and discuss the simulation results. Conclusion and perspectives are presented in Section 6.

2 Related Work

Several distributed MAC protocols have been designed for inter-vehicle communications. They can be classified into three categories, the contention-based medium access method CSMA/CA such as IEEE 802.11p [6], and the contention-free medium access method using Time Division Multiple Access TDMA, such as VeMAC [13, 14], TC-MAC [15]. The third category is a hybrid of the two previous methods such as DMMAC [17].

Recently, the TGp Task Group of IEEE [10] has proposed the IEEE 802.11p [6] to support VANET communications. Based IEEE 802.11, this standard tends to improve the QoS by using different messages priorities. In fact, IEEE 802.11p implements the Enhanced Distributed Channel Access EDCA [7] technique for packet prioritization. Nevertheless, a major problem of the IEEE 802.11p standard comes from the lack of bounded channel access delays guarantees [1], since it is based on a contention MAC method.

In [13] the authors propose a contention-free medium access control protocol for VANET called VeMAC. Vehicles in VeMAC are equipped with two radio interfaces, where the first is always tuned to the control channel CCH while the second one can be tuned to any service channel. CCH slot allocation is performed in a distributed manner where each vehicle randomly gets an available time slot. It is not the case for the SCH slot allocation. This task is ensured by service providers in a centralized way. However, packets transmitted by VeMAC on the CCH are large (Vehicle ID, current position, set of one-hop neighbours and the time slot used by each node within one-hop neighbourhood), which induces a high overhead on the CCH. Moreover, its random slot allocation technique is not efficient due to the appearance of free slots.

Günter and al. [14] propose a cluster based medium access control protocol (CBMAC). In their protocol each cluster head is responsible for time slot assignment to its cluster members for messages transmission. The aim of this protocol is to limit the effect of the hidden node problem and offer a fair medium access. In CBMAC protocol, the access time is divided into periodic frames and each frame is divided into time slots. The CH generates and manages the TDMA slot reservation schedule for its vehicles members according to the amount of data needed to send.

Recently a novel multi-channel MAC protocol called TDMA cluster-based MAC (TC-MAC) for Vehicular Ad hoc NETworks has been proposed by Almalag and al. [15]. It is based on stable clusters heads which are in charge of TDMA time slot assignment. TC-MAC provides an efficient time slots utilization for the exact number of active vehicles. Unlike WAVE MAC architecture, in TC-MAC protocol, the frame is not divided into two intervals CCHI and SCHI. In fact, each vehicle may switch to the Control Channel (CCH) or to a specific service channel (SCH) when needed during the time slot.

In [17], the authors introduce the Dedicated Multi-channel MAC (DMMAC) protocol. The DMMAC architecture is similar to WAVE MAC with the difference that in DMMAC, the CCH Interval is divided into an Adaptive Broadcast Frame (ABF) and a Contention-based Reservation Period (CRP). The ABF period composed of time slots, each time slot is dynamically reserved by an active vehicle as its Basic Channel (BCH) for collision-free delivery of the safety message or other control messages. The CRP employs CSMA/CA to organize its channel access. During the CRP, the vehicles negotiate and reserve the network resources on SCHs for non-safety applications. In addition, it is restricted to the only scenario of a straight highway road with an available number of slots higher than the maximum number of cars.

3 TDMA based MAC Protocol Challenges in VANETs Networks

The first aim of a MAC protocol for VANET is to ensure that each vehicle is granted an access to the channel in a bounded delay in order to send safety messages without collisions. TDMA is a method that can be used to assign one time slot to each active vehicle. We will study below the challenges of MAC solutions in VANETs focusing particularly on the TDMA technique. In the following we highlight the major problems faced in the case of a distributed TDMA slot allocation technique and in the case of a cluster based TDMA slot allocation technique.

3.1 Distributed TDMA Slot Allocation Challenges

When a distributed scheme is used to allocate a time slot, two types of collision on the time slots can happen [12] : access collision between vehicles trying to allocate the same available time slots, and merging collision between vehicles using the same time slots. The *access collision problem* [13] occurs when two or more vehicles within the same two-hop neighbourhood set attempt to allocate the same available time slot. This problem is likely to happen when a distributed scheme is used to allow the vehicle to reserve a time slot. As shown in Figure 1., the two vehicles attempt to access the same slot when they are within two-hops range. While *the merging collision* [12] is a basic problem for vehicular ad hoc network, this problem occurs when two vehicles in different two-hops sets using the same time slot become members of the same two-hop set due to their mobility. Figure 2 shown an example of the merging collision problem, when vehicle B in the first two-hop set is moving in opposite direction to vehicle E in the second two-hop set which is using the same time slot as B. Since B and E become members of the same two-hop set, then a collision will occur at vehicle D.

3.2 Centralized TDMA Slot Allocation

When the slot assignment schedule is centralized in the clusters heads, an inter-clusters interference problem can arise. There are two types of inter-cluster interference [4]: One Hop neighbouring Collision and Hidden Node Collision.

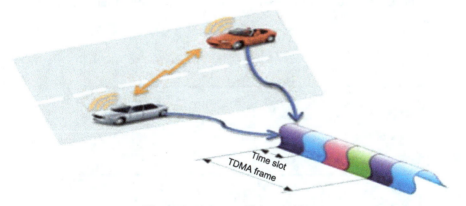

Figure 1   Access collision problem

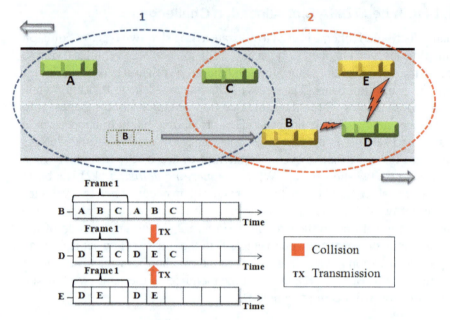

Figure 2   Merging collision problem

The ***One-Hop neighbouring Collision*** (OH-Collision) occurs when the same time slot is used by two neighbouring vehicles belonging to neighbouring clusters. Figure 3 shows an example of OH-collision situation, when vehicle C in cluster 1 and vehicle D in cluster 2 are using the same time slot. Since C and D are within transmission range of each other, then a collision will occur at vehicle C and D.

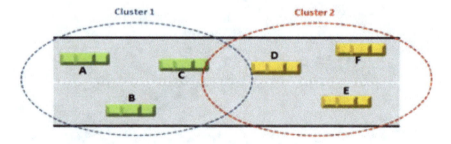

Figure 3   One-Hop neighbouring Collision (OH-Collision)

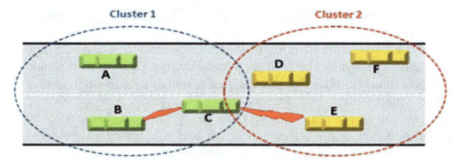

Figure 4   Hidden Node Collision (HN-Collision)

On the other hand, the ***Hidden Node Collision*** (HN-Collision) occurs when two vehicles are in range to communicate with another node, but not within transmission range of each other. Let us consider a situation in Figure 4 when vehicle B in cluster 1 and vehicle E in cluster 2 are using the same time slot. Since these two vehicles are outside transmission range of each other, a collision will occur at vehicle C in the cluster 1.

4 ASAS Protocol Description

ASAS strategy is based on a time division multiple access (TDMA) method, in which the medium is divided into frames and each frame, is divided into time slots. Only one vehicle is allowed to transmit in each time slot. This proposed strategy is centralized in stable cluster heads that continuously adapt to a high dynamic topology. The main idea is to take the direction and position of the vehicles into consideration in order to decide which slot should be occupied by which vehicle. The allocation of time slots is based on the requests from the vehicles in their HELLO messages, which are used by the cluster head to calculate the slot transmission schedule. The strategy is robust in the sense that it provides an efficient time slot reservation without intra-cluster and inter-cluster interferences. In this section, we address two important challenges: Cluster formation and the TDMA slot assignment mechanism for intra-cluster and inter-cluster communications.

4.1 Cluster Formation

Clustering technique is the process that consists to divide all vehicles in a network into organized groups called clusters. Several algorithms have been proposed for cluster formation that take into account the specific characteristics of VANET network such as [11] and [16]. We have proposed a cluster formation algorithm based on information of the vehicle's position and direction, and which uses the Euclidean distance to divide the vehicles into clusters. To provide more stable clusters, our cluster formation scheme takes into account the direction of vehicles, i.e. only vehicles moving in the same direction can be members of the same cluster. If the direction is not taken into account in a highway environment with two ways, the vehicles that are moving in opposite direction to the cluster head will only be part of the cluster for a very short time and a new cluster will have to be formed almost immediately. Through the Euclidean distance and transmission range (i.e. the DSRC range is 1km), we can decide whether two vehicles can be grouped in the same cluster.

Initially, all vehicles are in the Undecided State US. To divide the network into clusters, each vehicle broadcasts its state (direction, position and speed) to notify its presence to its one-hop neighbours. Then, based on the received messages each vehicle can build its one-hop neighbouring list. To determine the most stable CH, the elected cluster head is a vehicle which has the minimum average distance, the closet speed to the average speed and the maximum number of neighbouring vehicles. All the vehicles that are within transmission range of the elected CH become CMs and not allowed to participate in another cluster head election procedure. Once the clusters heads are elected, they maintain two sets of vehicles (see Figure 5): F (Front) and B (Back).

- B is a set of vehicles that are behind of the CH
- F is a set of vehicles that are ahead of the CH

Let C_i a cluster of size m and of cluster head CH_i defined by position (x, y, z).

- $F_i = \{V_{1 \leq i \leq m}(x', y', z'), x' \geq x\}$
- $B_i = \{V_{1 \leq i \leq m}(x', y', z'), x' < x\} = C_i - F_i$

After the cluster heads are elected as shown in Figure 5, each cluster head manages a local TDMA MAC frame. Moreover, after a cluster member CM receives its slot allocation from its cluster head, it transmits safety or control messages only during this slot and receives safety messages during the other time slots.

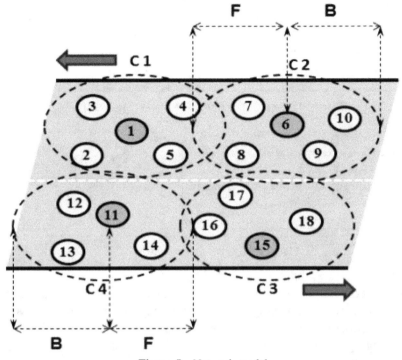

Figure 5   Network model

4.2 TDMA Slot Assignment in ASAS

In the scope of this work, we assume that each vehicle is equipped with a positioning system e.g. GPS (Global Positioning System) which can provide an accurate real-time three-dimensional position (latitude, longitude and altitude), direction, velocity and exact time. The synchronization among vehicles may be performed by using GPS timing information.

4.2.1 System Architecture

A vehicle is said to be moving in a left (right) direction if it is currently heading to any direction from north/south to west (east), as shown in Figure 6. Based on this definition, if two vehicles are moving in opposite directions on a two-way road, it is certain that one vehicle is moving in a left direction while the other vehicle is moving in a right one [13]. In ASAS, the channel access time is partitioned into frames and each frame is divided into ABS period and CRP period. The ABS period consists of a set of time slots where each time

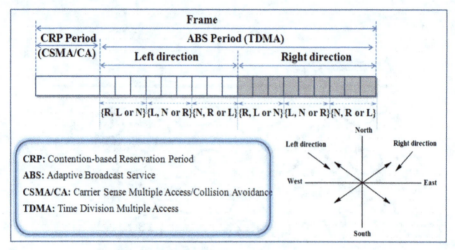

Figure 6   System Architecture

slot can be used by one vehicle only (CM, CG or CH) to broadcast a safety messages or other control messages such as topology management. The ABS period uses TDMA method as its channel access scheme. As defined in [13], to avoid merging collision problem, each ABS frame is divided into two sets of time slots, the first set is used by vehicles moving in left direction (see Figure 6) and the other is used by vehicles moving in right direction. The CRP period uses the contention-based method CSMA/CSA as its channel access scheme. During the CRP, if a vehicle needs to broadcast a message, it sends a request to the cluster head CH to reserve a periodic time slot. Then the CH responds to each reservation request from vehicle and assigns an available time slot in each ABS frame. We assume that each set of time slots Right or Left is partitioned into three subsets of time slots: *L, R* and *N*, as shown in Figure 6.

- L is the subset of time slots reserved for vehicles belong to the F set of vehicles.
- R is the subset of time slots reserved for vehicles belong to the B set of vehicles.
- N is the subset of unused time slots, in which all vehicles in the cluster remain inactive.

Moreover, to avoid the inter-cluster interference problem, the orders of the time slots subsets are different between neighbouring clusters. Hence, ASAS can reduce the inter- cluster and intra-cluster interferences without the use of other complex techniques such as CDMA, FDMA, SDMA and OFDMA.

4.2.2 TDMA slot reservation

In this section we provide a detailed description of our TDMA slot allocation strategy. When a vehicle V needs to access network, it first sends a reservation request to the cluster head CH for a periodic time slot. When CH receives the reservation request and depending on the vehicle position, it determines whether the current time slot belongs to the L, R set, and then it selects to V the first available slot as its owner slot. Each cluster head CH determines its distribution of three subsets of time slots "MAP" according to the MAPs of their neighbouring clusters. The CH can obtain the MAP information of the neighbouring clusters heads through the cluster gateways CG. Once a CH has selected a time slot for a CM, it sends a reservation which includes the slot identifier. However, ASAS requires that every CH should periodically send frame information FI to its two neighbouring clusters heads via its CGs. This information contains the following (see Figure 7):

1. CH-ID indicates the identifier of CH that sends the FI packet.
2. MAP {{R, L, N}; {L, N, R} or {N, R, L}}.
3. The sizes of R, L, N subsets.
4. The state of each time slot reserved for its moving direction.

The second information element is transmitted only once time and the third is transmitted if the cluster head updates the size of the *L, R* or *N* subsets. Unlike other slots reservation techniques based on FI broadcasts where each vehicle must determine the set of time slots used by all vehicles within its two-hop neighbourhood in order to acquire a time slot. In our reservation technique, the CH discovers the available slots while requiring less overhead than the others techniques. Moreover, the CH knows also all the time slots which likely to cause a collision at the transmission channel (i.e. N set). As shown in Figure 8, especially in the frame information of cluster head number 6 (FI-6), there is one available time slot for new vehicle moving behind of the cluster head. However, the reservation of any time slot which identifier belong to

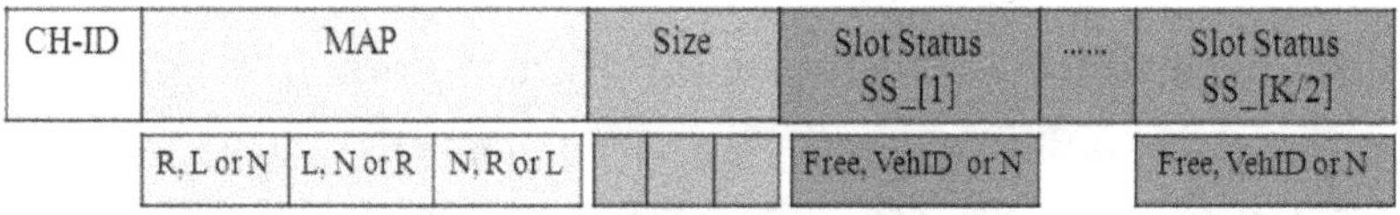

CH-ID	MAP			Size	Slot Status SS_[1]	......	Slot Status SS_[K/2]
	R, L or N	L, N or R	N, R or L			Free, VehID or N		Free, VehID or N

Figure 7   Frame information

Time Slot ID / Cluster ID	1	2	3	4	5	6	7	8	9	10	11	12	13	14	15	16	17	18
1										1	2	3	4	5		N	N	N
2										9	10		N	N	N	6	7	8
3	18			N	N	N	15	16	17									
4	11	12	13	14			N	N	N									

Figure 8   An example of slot assignment

[13 . . . 15] may cause a collision. When all the slots in L or R subsets are busy, the CH must communicate with its two neighbouring clusters heads to reserve a time slot in the N set for new vehicles respectively belonging to the F or B set.

The time slots are allocated according to the vehicle's movement and positions. By using a centralized approach we change the slot allocation process from random reservations to optimal allocations, which can improve the convergence performance of the MAC protocol and achieves an efficient broadcast service for a successful delivery of real-time safety information.

4.2.3 Release of TDMA Slot

If a cluster head does not receive a beacon message after a specific time from a CM to signal its presence, then the CH immediately releases the time slot allocated to that CM and it removes this CM from its cluster members list (i.e. the F or B set).

5 Performance Evaluation

In this section, we evaluate the effect of transmission range variation and we carry out a comparison of ASAS with the DMMAC and the VeMAC protocols. The simulation is based on an event-driven simulator implemented using Java language. We have used VanetMobiSim [8] to generate a mobility scenario.

5.1 Mobility Scenarios and Simulation Parameters

The mobility scenarios implemented for the highway are with two-way and different density levels (see Figure 5). The vehicles are moving at different speeds and have different transmission ranges. During simulation

Table 1   System parameters for simulation

Highway length	2km
Ways	2
Lanes/way	2
Transmission range/scenario	{150, 350, 550, 750, 1000}m
Slots/ Frame	50
Slots for left direction	25
Slots for left direction	25
Slot duration	1ms
Simulation time	120s
Number of vehicles/scenario	60
Speed mean value	100 km/h
Speed standard deviation	30 km/h

System parameters for simulation

time, each vehicle moves at a constant speed, and the number of vehicles on the highway remains constant. Table 1 summarizes the simulation parameters.

5.2 Performance Metrics and Simulation Results

We have evaluated ASAS based on the following performance metrics:

- MR-Collision rate: MR-Collision rate is defined as the average number of merging collisions.
- AC-Collision rate: The AC-Collision rate is computed as the average number of access collisions.
- IC-Collision rate: The IC-Collision rate is defined as the average number of inter-cluster collisions due to HN-Collision and OH-Collision. However for the DMMAC and the VeMAC protocols, the IC-collision rate is defined as the rate of collisions between the adjacent sets of two-hop neighboring vehicles that are moving in the same direction.

Due to the high dynamic topology, the number of clusters varies during the simulation time (new cluster are added and clusters are merged) and this variation should be as low as possible. Thus the cluster formation algorithm proposed reduces the number of new clusters created due to the high mobility of the vehicles. Moreover, it creates stable clusters and keeps the current clusters as stable as possible.

Figure 9, shows the rate of access collisions as function of different vehicle densities. Notice that there is no access collisions generated by ASAS, whereas both the DMMAC and the VeMAC protocols suffer from collisions. This can

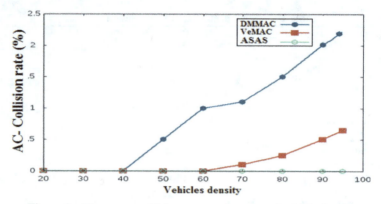

Figure 9   The access collision rate as a function of vehicle density

be explained by the centralized nature of the ASAS protocol, where cluster heads are in charge of time slots allocation. The DMMAC protocol as well as the VeMAC protocol generate higher rate of access collisions than ASAS, particularly for high traffic load. These results show the effectiveness of the ASAS technique.

Figure 10, shows the rate of merging collision for ASAS, VeMAC and DMMAC as function of transmission range. The ASAS protocol eliminates all merging collisions for the different transmission ranges studied. This comes from the fact that ASAS protocol assigns different sets of time slots to vehicles moving in opposite directions. The figure shows also that the merging collision rate is reduced by 100% compared to the DMMAC and VeMAC protocols.

In Figure 11 we depict the rate of IC-Collisions for the ASAS, VeMAC and DMMAC protocols. It is easy to see that ASAS have a lower rate of IC-Collisions than the two other protocols. The IC-Collision rate is reduced by 50% compared to VeMAC and by 5–15% compared to DMMAC. The main reason is that ASAS protocol allocates distinct sets of time slots to vehicles moving ahead and behind the cluster head. Consequently, ASAS protocol reduces collisions between neighboring clusters, which decreases the rate of Inter-cluster collisions compared to the DMMAC, and VeMAC protocols. We can also see that the IC-Collision rate decreases as the transmission range increases. This is because increasing the transmission range, decreases the number of clusters in the network and thus the inter-cluster collision rate will automatically decrease. We can conclude that ASAS protocol performs successfully under the DSRC architecture since the transmission range in DSRC is equal to 1000m. However, in case of low transmission range (less

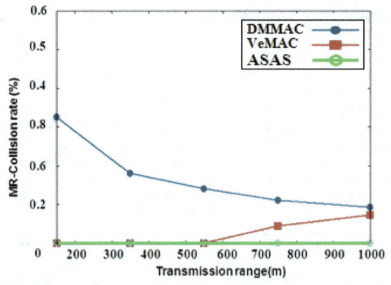

Figure 10 The merging collision rate as function of transmission range

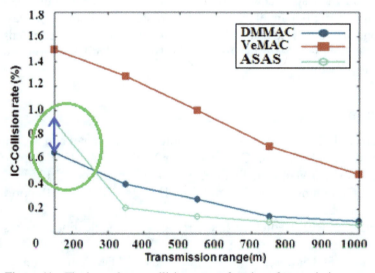

Figure 11 The inter-cluster collision rate as function of transmission range

than 250m) the DMMAC protocol presents better results than ASAS, because in this case, the large number of clusters increases the rate of inter-cluster collisions.

6 Conclusion and Future Work

This paper has studied the MAC protocol challenges to satisfy the requirements of real time and reliable broadcast of safety messages while achieving the fairness of the channel access. We have designed the Adaptive TDMA Slot Assignment Strategy (ASAS) to overcome these problems in which the assignment of time slots to vehicles is performed by the cluster heads in order to avoid any access collision problems. From the simulation results we conclude that this strategy achieves an efficient reservation and utilization of the available time slots without access collisions and decreases the rate of merging collisions as well as the rate of the inter-cluster collisions caused by the hidden node problem. Compared with the DMMAC and VeMAC protocols, ASAS generates a lower rate of transmission collisions in different transmission ranges and traffic load conditions. ASAS achieves this without having to use expensive spectrum management mechanisms such as CDMA or FDMA.

In future work, we will study the effect of various speeds and vehicles densities on the performance of ASAS. In addition, the dynamic adjustment of the length of the three subsets will be scrutinized. Moreover, we plan to extend ASAS to support multichannel operation and to provide reliable broadcast on both the control and service channels. In addition, we will carry out extensive simulations comparison with the IEEE 802.11p standard that operates with the DSRC architecture.

References

[1] K. Hafeez, L. Zhao and Z. Niu, Distributed Multichannel and Mobility-Aware Cluster-Based MAC Protocol for Vehicular Ad Hoc Networks, IEEE Trans. On Vehicular Technology, vol. 62, no. 8, oct 2013.

[2] Federal Communications Commission, FCC 99–305, FCC Report and Order, October 1999.

[3] The FCC DSRC (Dedicated Short Range Communications) web site. http://wireless.fcc.gov/services/its/dsrc/.

[4] T. Wu and S. Biswas, Reducing Inter-cluster TDMA Interference by Adaptive MAC Allocation in Sensor Networks, Sixth IEEE International Symposium on a World of Wireless Mobile and Multimedia Networks (WoW-MoM'05), 2005.

[5] CAR 2 CAR Communication Consortium. http://www.car-to-car.org/.

[6] 802.11p-2010-IEEE standard for information technology-Telecommuni cations and information exchange between systems-local and metropoli-tan area networks – specific requirements part 11 : Wireless LAN medium access control (MAC) and physical layer (PHY) and physical layer (PHY) specifications amendment 6 : Wireless access in vehicular environments. 2010.

[7] 802.11–2007 - The institute of electrical and electronics engineers IEEE standard for information technology – telecommunications and informa-tion exchange between systems - local and metropoli-tan area networks - specific requirements. Part 11: Wireless LAN medium access control (MAC) and physical layer (PHY) specifications.

[8] VanetMobiSim project, home page (2010). http://vanet.eurecom.fr (accessed: May 29, 2010).

[9] Q. Tse, Improving Message Reception in VANETs, in Proc. of Inter-national Conference on Mobile Systems, Applications and Services (MobiSys), June 2009.

[10] TGp. http://www.ieee802.org/11/Reports/tgp update.htm.

[11] P. Fan, P. Sistla and P. C. Nelson, Theoretical analysis of a direc-tional stability-based clustering algorithm for vanets, Vehicular Ad Hoc Networks, 2008.

[12] F. Borgonovo, L. Campelli, M. Cesana and L. Fratta, Impact of user mobility on the broadcast service efficiency of the ADHOC MAC protocol, Proc. IEEE VTC, vol. 4, pp. 2310–2314, 2005.

[13] H. A. Omar, W. Zhuang and L. Li, VeMAC: A TDMA-Based MAC Protocol for Reliable Broadcast in VANETs, IEEE TRANSACTIONS ON MOBILE COMPUTING, VOL. 12, NO. 9, September 2013.

[14] Y. Günter, B. Wiegel, and H. Grossmann, Cluster-based medium access scheme for vanets, Intelligent Transportation Systems Conference, 2007. ITSC 2007. IEEE, pp. 343–348, Oct 2007.

[15] M.S. Almalag, S. Olariu and M.C. Weigle, TDMA cluster-based MAC for VANETs (TC-MAC), Proc. IEEE WoWMoM, pp. 1–6, 2012.

[16] C. Shea, B. Hassanabadi and S. Valaee, Mobility-based clustering in VANETs using affinity propagation, in IEEE Globecom, 2009.

[17] F. Liu N. Lu, Y. Ji and X. Wang, DMMAC : A dedicated multi-channel MAC protocol design for VANET with adaptive broadcasting, in Wireless Communications and Networking Conference (WCNC), 1–6, Sydney, Australia, 2010.

Biographies

Mohamed HADDED received Bachelor's Degree in Computer Science from the Faculty of Science of Gabes and MS degree in Computer Sciences and Information Systems from the Higher Institute of Computer Science and Mathematics. He is currently a PhD student at the National School of Computer Science of Manouba. His research interests include vehicular communications, mobility management, and Quality of Service.

Dr. Rachid Zagrouba is an assistant professor from September 2008 at university of Tunis El Manar (Tunisia). He received his Computer Science Engineering and Master degrees from the National School of Computer Science (ENSI) in 2001 and 2002, respectively. At the end of 2003, he joined the RSM research team at Telecom Bretagne in Rennes (France) to prepare a Ph.D. in Computer Science which he has defended on December 2007. During 2006 and 2007, he was an ATER (Attach Temporaire d'Enseignement et de Recherche) at the University of Rennes 1. From September 2003 to September 2006, he was with the Mobile Communications team of orange Lab in Rennes as a research engineer. He was involved in several French-funded and IST FP6/7 projects. He served as a technical reviewer of several international conferences and journals. He is the Ph.D. Supervisor of three Ph.D. students in the area of computer networking, wireless and cellular networks, vehicular communications, and mobility management.

Dr. Anis Laouiti received his PhD in computer science from the Versailles University, France, in 2002. He had been doing his research during and after his Phd at INRIA/Hipercom team, before he joined the TELECOM Sud-Paris, France, as an associate professor in 2006. His research interests include unicast/multicast routing protocols for MANET, and vehicle to vehicle communications. He was involved in the IETF-MANET working group and he is one of the co-authors of the OLSR routing protocol.

Pr. Paul Muhlethaler was born in february 1961. He graduated from Ecole Polytechnique in 1984. He received his PhD in 1989 from Paris Dauphine university and its habilitation in 1998. He is researcher at INRIA since 1988. He is now research director at INRIA where he co-founded the HiPERCOM team with Philippe Jacquet. His research topics are mainly around protocols for networks with a speciality in wireless networks. He had also a few, often referenced, results on scheduling issues. In wireless networks, he has actively worked at ETSI and IETF for the HiPERLAN and OLSR standards. He is now following the European standardization for vehicular networks. Another important aspect of his activity concerns models and performance evaluations. He was the first to carry out optimizations of CSMA protocols in Multihop Ad Hoc Networks. With F. Baccelli and B. Blaszczyszyn he derived a complete model of an Aloha multihop ad hoc net-work. Interestingly this model lead to the design of one of the first multihop ad hoc network offering a throughput scaling according the Guta and Kumar's famous law. He is now particularly interested in deeply understanding the achievable performance of multihop ad hoc networks and in tracking all the possible optimization ways of such networks. In 2004, he received the pres-tigious price "Science et Dfense" for his work on Mobile Ad Hoc Networks. He was among the first contributors for the OLSR protocol in 1997.

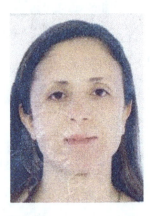

Pr. Leila Azouz Saïdane is Professor at the National School of Computer Science (ENSI), at The University of Manouba, Tunisia and the Chair person of the PhD Commission at ENSI. She was the Director of this school and the supervisor of the Masters Degree program in Networks and Multimedia Systems. She is the head of RAMSIS group of CRISTAL Research Laboratory (Center of Research in Network and System Architecture, Multimedia and Image Processing) at ENSI. She collaborated on several international projects. She is author and co-author of several papers in refereed journals, magazines and international conferences.

www.ingramcontent.com/pod-product-compliance
Lightning Source LLC
LaVergne TN
LVHW012332060326
832902LV00011B/1849